John Henry Steel

Outlines of Equine Anatomy

a manual for the use of veterinary students in the dissecting room

John Henry Steel

Outlines of Equine Anatomy
a manual for the use of veterinary students in the dissecting room

ISBN/EAN: 9783337825287

Printed in Europe, USA, Canada, Australia, Japan

Cover: Foto ©Andreas Hilbeck / pixelio.de

More available books at **www.hansebooks.com**

OUTLINES

OF

EQUINE ANATOMY

A MANUAL

FOR THE USE OF VETERINARY STUDENTS IN THE

DISSECTING ROOM.

BY

JOHN HENRY STEEL, M.R.C.V.S.,

ASSISTANT-DEMONSTRATOR OF ANATOMY, ROYAL VETERINARY COLLEGE.

NEW YORK:
WILLIAM WOOD AND CO., PUBLISHERS.
27, GREAT JONES STREET.

PREFACE.

THIS sketch of the Structure of the Horse is but an *outline;* the same subject in different aspects has been treated by CHAUVEAU (rendered available to us by FLEMING's Translation), and by GAMGEE and LAW ; and the manner in which it has been executed by them, in both cases, betrays the hand and touch of the master. In those works, however, the outline has been filled up, and into the background have been introduced figures of other domesticated animals, the beauty of which draws the attention away from the main object. The anatomical works of Professor Strangeways and of Percivall are fully appreciated by the Author, but he has found that a concise work with an arrangement illustrating the sequence of structures in the course of dissection is much required to introduce the student to more complete and pretentious records.

A tribute of respect to Professor Pritchard the Author feels to be a duty; his sound anatomical knowledge, and his skill and eloquence in communication of information to those who have the pleasure of studying under him, have indelibly impressed the mind of the Author as of numerous other members of the veterinary profession.

The Author has hoped, therefore, by these outlines to introduce the student of veterinary comparative anatomy to his standard of comparison, for only by first gaining a knowledge of this can he qualify himself for the study. In the dissecting room the student's investigations are in the first place confined to the Horse, afterwards he may avail himself of opportunities of examining other animals. It is, therefore, in the dissecting room, during this first stage of his study, this work is designed to render him assistance. In his study and at his leisure he will fully appreciate the beauties and the solid value of the works above alluded to ; but the student in the dissecting room requires CONCISENESS, a quality which the Author believes he has been successful in harmonising with TRUTH.

And if, in the course of his examination of *Nature*, the student observes errors in this work of *art*, let him remember "Varium et mutabile semper natura;" let him see whether he has not discovered one of those exceptions which but serve to prove Nature's rules.

A work of this kind necessarily omits important facts, but the author has endeavoured to insert all the leading facts of his subject, almost all of which he has himself seen in the dissecting room. When obliged to trust to the observations of others, he has referred to our leading authorities; he has not, however, hesitated to confirm his own observations by comparison with those recorded by others. If he has failed in supplying to veterinary literature a work of which he himself not very long ago, as a student, felt the need, he hopes his failure will stir up some abler hand to the work; and that the boldness of this attempt will be pardoned in consideration that the aim of the author is to facilitate the acquisition of a knowledge of Veterinary Anatomy, and thus slightly to serve that profession to which it is at once his pride and his honour to belong.

J. H. S.

Royal Veterinary College;
Nov., 1876.

OUTLINES

OF

EQUINE ANATOMY.

PART I.—INTRODUCTORY.

ANATOMY is the examination by means of dissection (cutting up) of organized bodies, and the scientific arrangement and appreciation of the parts thus separated with regard to their character and relative position.

The science of anatomy is therefore very extensive; it is primarily divided into animal and vegetable anatomy; and animal anatomy is called **Anthropotomy** when man is the object of investigation, **Hippotomy** when the horse is being examined. In the study of **Comparative Anatomy** some particular species of animal is taken as the standard, and the points in which other species differ from this are noted: *it is the search for diversity.* **Transcendental or Philosophical Anatomy** is the search for analogies between different species and the endeavour to reduce them all to a general type.

Investigation demonstrates that all the organs of the body are made up by the various combinations of a few primary tissues produced by union of anatomical elements. The study of these tissues, irrespective of their situation, is termed **General Anatomy**, while the investigation of their component elements by the aid of the microscope is **Histology**.

An **organ** of the body is a collection of definitely arranged tissues for the performance of some function useful to the animal economy; the lowest forms of animal life consist of but one organ, a simple cell; but in ascending the animal scale we find organ after organ superadded, until in the vertebrata we find a most complex collection of mutually dependent organs; the study of the characters and relative positions of these and the arrangements of the tissues of which they are composed is termed **Descriptive Anatomy.**

The examination of healthy organs is **Physiological**, of diseased, morbid or **Pathological Anatomy.** A knowledge of anatomy enables the physician to diagnose (distinguish) disease; the surgeon to diagnose, appreciate the extent of, and apply the best means of remedy for injuries and malformations; and the natural historian to arrange the various beings, and so to suggest those most fit for economic purposes. An anatomist only can duly appreciate the importance of the ever varying points constituting " unsoundness " in the domesticated animals.

The **instruments** required in the pursuit of descriptive anatomy of the horse are not numerous. *Scalpels* are the agents used in the separation of the soft organs one from the other, a firm grasp being maintained upon the most pliable by means of *forceps*, while, if necessary, the opposing tissue is retained in position by *hooks and chains*. The cutting instruments require to be kept sharp by means of an oil-stone. In some cases *scissors* may be substituted for scalpels with advantage, as in clearing arteries and in dissecting the eye. A *blowpipe* is a useful accessory for the inflation of hollow organs with thin walls; a *saw* is absolutely necessary for the proper demonstration of the various bony cavities and of the structures contained in them, while a *chisel and hammer* serve to raise the separated portions of bone.

The vessels require to be filled with some substance which will distend them as they naturally exist in the living animal, and thus render them more apparent. Arteries and veins for temporary use are generally distended with coloured size or tallow, which solidify as the body cools, but more permanent specimens require coloured wax; mercury is useful for injecting lymphatics. The substance is forced into the vessels by means of an *injecting syringe* which has *stopcocks*. These stopcocks require to be introduced into the large arterial trunks, when the material will fill the arteries of the body, but the valves of the veins prevent fluid from passing from their larger to their smaller branches, so it is here necessary to commence at the smaller vessels. Lymphatics are injected by introducing the point of the syringe into the areolar tissue beside the arteries and veins of an organ, when the mercury will generally find its way to the lymphatics; they are, however, but seldom dissected, and being valvular like the veins, cannot be injected from the larger vessels.

Solutions of nitric and other acids and alcohol are used for hardening the softer organs, as the brain and the eye, while by *maceration* or *boiling* the bones may be cleaned, and thus rendered fit for demonstration of their peculiarities; by boiling also we render apparent the arrangement of the muscular fasciculi of the heart. The *age of the subject* influences its value for dissecting purposes. Young animals, in consequence of the tenderness of their areolar tissue, do not present that similarity between bands of white fibrous tissue and the nerves which we find in the older subject. They also present well-marked separations between the bones, and in many cases the various ossific centres from which these bones originate; the muscles, though more easily separable, are less marked and more liable to tear; an older subject is, therefore, preferable for dissection of muscles. Fat subjects are said to keep better than lean, but this advantage is more than counterbalanced by the difficulty of removal of the adipose tissue from some parts. Coarse lymphatic animals present bulky but soft muscles, while well-bred horses and asses afford firm well-defined muscles. Since also they occupy less space and are less expensive, asses are generally dissected by the veterinary student. They do not differ *essentially* in structure from the horse. The value of an animal for dissection may, after death, be judged by taking into consideration the period before accession of rigor mortis or death stiffening and the length of its duration. When death has ensued from nervous prostration the vital powers soon lose their control over the tissues. Coagulation of the albuminoid matter in the muscles occurs, constituting rigor mortis, but soon decomposition commences as a result of the substitution of chemical for vital force in the tissues. Such a subject, therefore, is liable to decompose sooner than one of opposite characteristics. Human bodies are preserved by injection of the vessels with solution of arsenic, carbolic acid, and other preservative fluids, but such a precaution is not necessary in veterinary anatomy, since our subjects are destroyed when in a state of health, are cheaper, and will keep for three weeks with care. The bodies of animals which have died or been destroyed when suffering from disease are deficient in tone or firmness of the tissues in general. Death from hæmorrhage or bleeding causes a general want of colour of the parts, and therefore of one of the most useful guides in separation, while puncture of the brain by the poleaxe or "pithing" produces satisfactory results with instant death.

When placed on the dissecting-room table the subject is generally rested on its right or left side for the convenience of the dissectors, six of whom work at the same subject, dividing it among themselves into six parts—head, neck, abdomen (including the pelvis), back (including the thorax), fore and hind extremities. Such a division, though extremely artificial, is advantageous, though it requires cordial co-operation at the boundaries of the various regions; thus, the longus colli lies for the most part in the neck, but extends into the thorax; the proper demonstration of this muscle, therefore, requires the action of two individuals. The subject should be placed with the abdomen towards the light, as the organs contained in it are of much importance and require careful dissection. We may here observe that some art is required in the distinction of the different structures in different positions of the subject; diversity in this respect, therefore, while it renders the student proficient in practical anatomical knowledge, but ill accords with that "cramming" for examination which consists in the retention *verbatim* of passages from books, and the fixture of pictures of the subject in definite positions on the mind.

The *primary lines of incision through the skin are*—
    I. From the root of the ear to the throat just behind the angle of the jaw.
   II. From the withers to the point of the shoulder and on to the centre of the chest.
  III. From the commencement of the back to the under surface of the chest near the elbow.

IV. From the posterior part of the loins to the central line of the abdomen against the stifle.
V. From the centre of line No. III to the centre of line No. IV.
VI. Along the central line of the under surface of the body from the lower lip to the anus.

In making these incisions care must be taken to cut the skin only, for under it in some places important structures are superficially situated. The skin having been punctured by a slight incision at first, the knife (this work takes the edge off an ordinary scalpel) should be introduced, and the remainder of the skin divided from within outwards. From the primary incisions the dissectors then commence to separate the skin from the neighbouring parts, being careful to cut only through areolar tissue, the great connecting medium of the other tissues; the free flaps in a short time produced may be then drawn away from the subject, whereby the areolar tissue is rendered tense and thus more easily perceptible and divisible. Any blood effused by accidental or intentional incision into veins is to be removed with a sponge, and only such parts as are to be immediately dissected are to be denuded of skin, for on exposure the areolar tissue becomes dry and tough, assuming the appearance of dried ligament or tendon in some places, and hence becoming liable to mislead. All portions of separated skin should be immediately removed, as being liable to cover the subject with loose hairs. The bare portions may be kept moist by damp cloths, which also serve to keep off dust, &c. After one side of the animal has been duly investigated much information may be gained by turning it over and examining the other side by a similar but somewhat modified process, each dissector exchanging his region for some other, whereby the most satisfactory amount of information gained from each subject by each individual is attained.

Finally, we will recommend that, while acquiring that superficial knowledge of the situation and peculiarities of the bones which is requisite to the due appreciation of the anatomical position of organs, the student avail himself of the opportunity of learning practically the handling of the scalpels, skinning, cleaning, &c., *for the first subject of a student generally yields him but little more information than this.* Never cut through anything but areolar tissue without having examined its nature. In the separation of the skin the edge of the knife must be always turned towards the skin from the unknown tissues, this is a primary rule in dissection; always cut towards known structures when dividing areolar tissue, any slip of the knife will then do no damage.

Where possible, have the basement structure of the part, as the bones or cartilages, cleaned to show its points of muscular attachment, &c., near at hand for constant reference, and during dissection take advantage of the guidance of some authentic work on anatomy read aloud, or better, of a good anatomist in *propriâ personâ*. Carefully note any point in which the subject under examination differs from the description.

The different portions of the subject may be altered in position, irrespective of the position of the subject in general, by means of *blocks* placed under parts which require to be elevated, *iron-forked bars* firmly fixed in the table by penetration with one sharp extremity, which form a prop for limbs, and *cords* either from pulleys or from fixed points, these may be tightened at convenience.

By the separation of organs one from the other in the dead body the hand and the eye are trained for operations upon the living subject; the student therefore should not regret the time expended in the mechanical pursuit of anatomy, nor consider that examination of preparations, plates, or recent specimens prepared by others, will render him proficient in anatomy; it is only men who work on this system who fear the *practical* anatomical examination recommended as a test by the highest authorities.

One of the most prominent characteristics of the highest subkingdom of animals is the possession of an internal skeleton or basement structure. This consists of a number of definitely but diversely shaped masses of a hard white substance composed of a mixture of earthy and animal matter. Submit a bone to combustion in a fire with a free supply of air, it first turns black in consequence of the carbon contained in its animal matter, and when this is completely burnt a white friable mass, in shape resembling the bone, remains, being the earthy matter which entered into its formation. Subject another bone to the action of dilute hydrochloric acid, this will remove the earthy matter, and the animal matter will remain in the shape of the bone, but extremely flexible.

In the extremely young subject all the bones are cartilaginous, being elastic and pliable, but afterwards, by the deposit of earthy matter in their substance and by alteration in the arrangement of their component elements, they change to bone. This process of **ossification (bone making)**, commencing at fixed points, involves gradually the surrounding cartilage. These points are termed **centres of ossification**. Some cartilages remain of fixed structure throughout life, and are termed "*permanent.*" They with the bones therefore form the **skeleton**, which is composed of many portions brought into movable contact with each other, whereby the various motions of the body are possible and concussion from motion modified, and thus shock to internal organs prevented. Where firm power of support and resistance is required bone is situated; where yielding with protection to other parts, cartilage. When these structures are placed in position and maintained there by artificial means we have an *artificial skeleton*, but in the *natural skeleton* they are bound together by bands or layers of a strong glistening fibrous substance (*white fibrous tissue*) or of a *yellow elastic tissue*, the characters of which are expressed by its name. These are *ligaments*. The **white fibrous tissue** under the microscope is found to be composed of collections of parallel bands of wavy fibres with a single outline, which bands cross each other irregularly. **Yellow elastic tissue** consists of fibres with a double outline, which are irregularly arranged, but send branches to the neighbouring fibres, and are curled at their extremities in consequence of their elastic recoil.

Cartilage consists essentially of *cells* (*circumscribed sacs, each containing smaller sacs or nuclei, which contain a central spot or nucleolus*) with an *intercellular substance or matrix*. In the **true or hyaline cartilage** this matrix is devoid of structure, being slightly granular; but when cartilage requires increased strength we find the matrix assuming the characters of white fibrous tissue; where it requires elasticity, yellow elastic tissue; thus, **white fibrocartilage** and **yellow elastic fibro-cartilage** are formed. These three forms of permanent cartilage present cells collected in groups of about three or four, which have been produced by division of one cell, but the cells in **temporary cartilage** are more evenly diffused through the mass. Another form of cartilage is found in the fœtus, which is termed **cellular**, since it consists of cells collected together with no apparent intercellular substance.

Bones are composed of *compact* and *cancellated structure*. The former is the dense external layer which they present. In the cancellated, which is situated more internally, the bony structure is arranged in a trabeculated manner, whereby spaces are left which are termed cancelli. Small near the compact substance, these gradually increase in size until in the centre of the bone we frequently find a large open space or *medullary canal*, so named from its containing the **medulla or marrow**, a greasy substance, which in the young subject is of a red colour, but in the adult yellow; it consists of areolar tissue, in the meshes of which are fat and *peculiar cells* (or *myeloplaxes*); it also fills the cancelli.

The **intimate structure of bones** is complex; through its substance run canals for blood-vessels, and arranged concentrically around these canals are bony layers termed *laminæ*, between which at intervals are irregular spaces containing the special cells of bone, or *lacunæ*. From each of these lacunæ, towards the central or *Haversian canal*, and towards the other lacunæ, run minute passages or *canaliculi*, through which only the fluid portion or nutritive plasma of the blood is supposed to pass. An arrangement such as that just described is termed an **Haversian system**, and the Haversian canal of one system is frequently brought into communication with that of another by means of a transverse branch, all the Haversian systems composing a bone are not uniform in size.

Bones are of three kinds, long, flat, and irregular. **Long bones (or long round bones)**, as the femur, radius, &c., have one axis (as a rule) much longer than the others and have a distinct medullary canal. They present a *central ossific centre or diaphysis* forming the body of the bone, and at each end of this an *epiphysis* forming an extremity. From either part *processes or apophyses* may project. Their Haversian canals run parallel to the long axis of the bone, and the different Haversian systems are connected together by connecting laminæ, which are concentric with the medullary canal of the bone.

Flat bones have one axis much shorter than the others, whence the bone assumes an expanded form and serves for the attachment of large muscles or for forming the walls of cavities. They present no medullary canal, but between the outer compact layers is cancellated substance, which in the cranial bones is termed diploë. Examples of flat bones —scapula, parietal bones. They have a centrally situated centre of ossification, with others for their processes, and their Haversian canals are perpendicular to their flat surfaces. Some parts of them seem to result from ossific deposit in fibrous membrane, and not from change of cartilage.

Irregular bones are those in which no marked difference exists between the length of their axes. They consist of cancellated structure surrounded by a thin layer of compact substance. The centres of ossification are related to the peculiarities of each bone. Examples—petrous, temporal bones, os pedis, &c. In this class are included "*short bones.*"

Bones derive their supply of blood from the arteries which run nearest to them; they are surrounded by a strong membrane composed of white fibrous tissue, **periosteum**, which is very vascular, and from which minute vessels pass to foramina on the surface of the bone in connection with the Haversian canals. In addition to this long bones generally present a small opening (*foramen medullare*), which pierces the shaft of the bone, and through which passes a branch from the nearest important artery which runs to the medullary canal, and breaks up to supply the medulla and to send small branches, corresponding with those from the periosteum, into the bone; and a third source of supply is derived from small *articulatory foramina* surrounding those surfaces by which these bones come into indirect contact with each other. These articulatory surfaces being covered with cartilage, are not covered by periosteum, which also is absent from some places where tendons become directly attached to the bone, and from bones, like the ethmoidal cells and the ossa turbinata, which are intimately clothed with highly vascular mucous membrane. Bones are but slightly supplied with nerve-force, their lymphatics have not been satisfactorily demonstrated. Bones present **eminences** and **cavities**, either of which may be *articulatory* or *non-articulatory*.

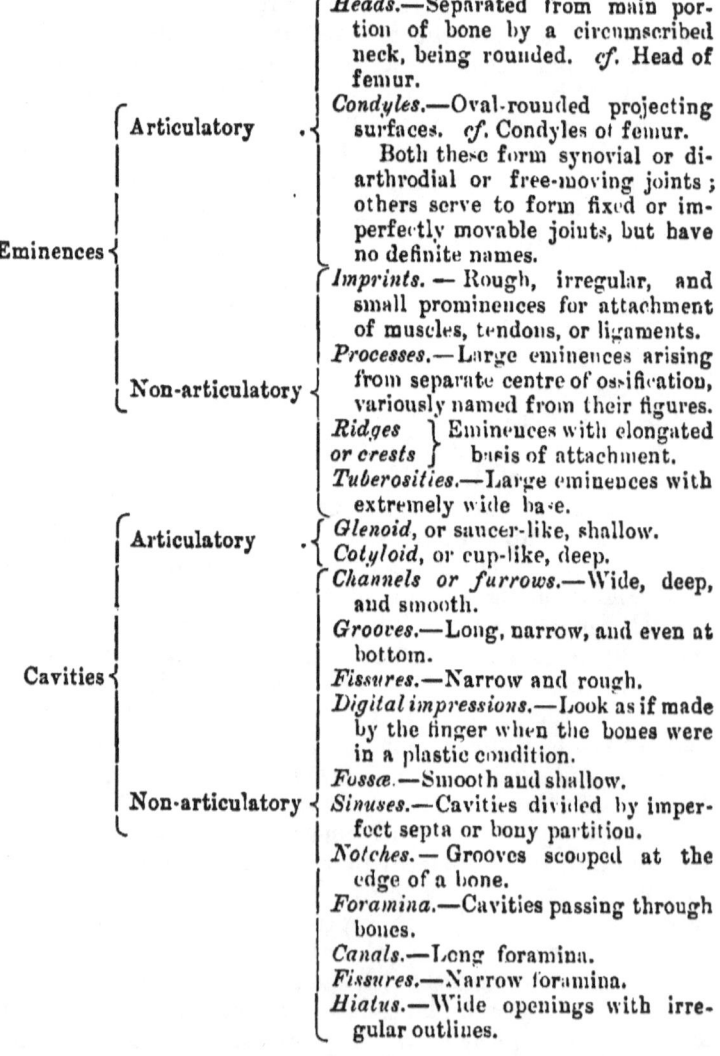

- **Eminences**
  - **Articulatory**
    - *Heads.*—Separated from main portion of bone by a circumscribed neck, being rounded. *cf.* Head of femur.
    - *Condyles.*—Oval-rounded projecting surfaces. *cf.* Condyles of femur.
    - Both these form synovial or diarthrodial or free-moving joints; others serve to form fixed or imperfectly movable joints, but have no definite names.
  - **Non-articulatory**
    - *Imprints.* — Rough, irregular, and small prominences for attachment of muscles, tendons, or ligaments.
    - *Processes.*—Large eminences arising from separate centre of ossification, variously named from their figures.
    - *Ridges or crests*} Eminences with elongated basis of attachment.
    - *Tuberosities.*—Large eminences with extremely wide base.
- **Cavities**
  - **Articulatory**
    - *Glenoid,* or saucer-like, shallow.
    - *Cotyloid,* or cup-like, deep.
  - **Non-articulatory**
    - *Channels or furrows.*—Wide, deep, and smooth.
    - *Grooves.*—Long, narrow, and even at bottom.
    - *Fissures.*—Narrow and rough.
    - *Digital impressions.*—Look as if made by the finger when the bones were in a plastic condition.
    - *Fossæ.*—Smooth and shallow.
    - *Sinuses.*—Cavities divided by imperfect septa or bony partition.
    - *Notches.*— Grooves scooped at the edge of a bone.
    - *Foramina.*—Cavities passing through bones.
    - *Canals.*—Long foramina.
    - *Fissures.*—Narrow foramina.
    - *Hiatus.*—Wide openings with irregular outlines.

The epiphyses of long bones consist mainly of cancellated tissue, the compact portion forming but a thin external layer; that under the cartilage (*articular lamina*) is most dense, and from its vessels the cartilage-cells draw nutriment. This enlargement of the extremities of the bone is for the purpose of affording increased surface for ligamentous and other attachments without increasing the weight of the bone.

For convenience of description anatomists, when possible, distinguish *surfaces of bones*, and the lines of union of the surfaces constitute the *borders*.

We have mentioned **ligaments** of various kinds entering into the formation or perfection of joints. These are generally composed mainly of white fibrous tissue; some contain a small amount of yellow elastic tissue; others *mainly* consist of yellow elastic tissue, as ligamentum nuchæ. We have seen, too, that the ligaments are either broad expansions or dense bands. Analogous in structure are the **fasciæ**, which are found in particular regions; they consist of widespread membranous layers of white fibrous tissue, which form sheaths to bind down muscles in their situations and to brace them for action; they are generally attached to the most prominent portions of the basement structures in the neighbourhood. They present channels and foramina through which vessels and nerves pass, and in some cases have muscles acting upon them, as tensor vaginæ femoris. They are most marked in those situations in which the muscles are long, as in the limbs. On approaching an annular or a capsular ligament they blend with it. The most important are fascia lata of the forearm, the lumbar, psoas, and gluteal fasciæ, and the fascia of the thigh. The gluteal presents a considerable amount of yellow elastic tissue. Fascia superficialis abdominis is wholly elastic. The several portions of the skeleton are caused to move upon each other by the action of **muscles**. The motion found in other parts of the body, with the single exception of that produced by ciliæ, is also due to the peculiar property of muscular fibres; this motion may be *voluntary* or *uncontrolled by the will*, and is brought about by the contractility of two forms of fibres, *voluntary or striated, involuntary or unstriated*. **Involuntary muscular fibre** is that which produces the motion of most of the internal organs by which nutrition and reproduction are brought about; it is generally arranged so as to form a thin coating to a hollow organ or a tube or a vesicle, and consists of a number of fusiform or spindle shaped nucleated cells overlapping each other at their extremities where they meet. The double wall of separation becomes absorbed and thus a plain fibre is produced, distended at intervals in consequence of the presence of the nucleus; these fibres are collected together to form fasciculi, which are generally spread out, forming layers, but sometimes, as in the large intestine, are collected to form bands. When in mass they are white in colour. **Voluntary, striated, or red muscular fibre** is that which is under the control of the will under ordinary circumstances. It consists primarily of certain minute globular cells, *sarcous elements*, which are arranged so as to form elongated chains or *fibrillæ*. When examined under any but the highest powers of the microscope the lines of junction of the cells seem like black lines or *striæ* extending across the fibrillæ. A number of fibrillæ become collected into a bundle, surrounded and connected together by a layer of simple membrane, the *sarcolemma*, to form a *fibre;* and the fibres are collected into *fasciculi*, which vary in size, may be seen with the naked eye, and are surrounded by *perimysium*. These fasciculi are united together to form masses of various shapes, red in colour, vulgarly known as "flesh." These are the **muscles** of anatomists; the study of them constitutes myology. Muscles vary much in form, being in some cases fusiform, in others flat, and in others irregular; they are attached to the basement structures in most cases, and sometimes consist wholly of red muscular fibre, in others the red muscular portion, *belly*, is continued by a white ligamentous band, which may be cordiform, or expanded and flat (aponeurotic). This band is termed a **tendon**; it consists of dense, slightly modified, white fibrous tissue, and in some places becomes firmly blended with the periosteum of the bones to which it runs, in others becomes continuous with ligaments. When the muscular fibres of a muscle meet the tendon in a direction obliquely to its course on one side the muscle is termed *penniform;* if on both sides, *bipenniform;* sometimes two fleshy bellies of a muscle are connected by a central tendon, such a muscle is said to be *digastric*. Muscles, as a rule, cover the surfaces of bones and form grooves and channels through which the nerves and blood-vessels pass, and as they are marked more or less distinctly in different cases by prominences or depressions on the surface of the body of the living animal, they afford valuable surgical guides. Vessels and nerves which run in contact with a muscle are termed its *satellites*. The muscles are extremely numerous, they vary in size from the huge longissimus dorsi to the minute muscles of the middle ear. The tendons of some muscles extend throughout their whole length, performing the function of a ligament in addition to their motorial value. Muscles are generally connected together by areolar tissue, in dissecting them the knife should be always used in the direction of their fibres, their attachments more particularly investigated, but their relations and figure also taken into account.

Nerve-tissue next demands notice, as being the source of all muscular action, sensation, and intellect. The nervous system is primarily divided into the **cerebro-spinal** and **sympathetic systems**, each of which consists of nerves and nerve-centres; the two are intimately blended. The **cerebro-spinal nervous system** is that which regulates animal life, *i. e.* sensation, voluntary motion, and intellect. Its *centres* are the *brain and spinal cord;* its nerves are marked by whiteness, they are given off from the above-mentioned centres, as a rule, in pairs, one from each side of the centre. *Those of the spinal cord originate by two sets of fibres,* a **superior or sensory,** from the supero-lateral part of the cord; each of these has a ganglion upon it and subsequently unites with the **inferior or motor root.** After this the sensory cannot be distinguished in any way from the motor fibres; the difference lies not in the fibres, but in the parts to and from which they run. The **sympathetic, ganglionic, or vegetative system** superintends those organs which provide directly for the maintenance of the individual or the perpetuation of the species. Its *centra* consist of a number of grey nodular bodies, enlargements of the nerves, found in various parts of the body (termed *ganglia*). Thus, they exist as a chain of ganglia connected together by fibres on either side of the spine, as special ganglia existing in certain organs as the heart, as the ganglia on the sensory roots of the cerebro-spinal nerves, and as certain ganglia situated in certain irregular but constant parts of the body, as those of the fifth cranial nerve. From these fibres run, collected into nerves distinguishable by their darker colour than those of the animal system. These and the cerebro-spinal nerves communicate in every possible manner. *Two classes of histological elements* are distinguishable in nerve-tissue, *fibres and cells*. The nerves are composed only of **fibres** which are of *two kinds, grey and white*. The *grey fibres* so much resemble unstriated muscular fibre that their nervous nature has been doubted. They are found most largely in the sympathetic nerves, of which they form the greater part, and consist of transparent fibres marked by a number of nuclei; they are smaller in calibre and their nuclei more crowded than those of unstriated muscular fibre. The *white fibres* in the living and very fresh subject are transparent, resembling extremely minute glass threads, but shortly after death they undergo change and exhibit a central transparent but tough substance, *axis cylinder*, surrounded by a thin layer of simple membrane, *medullary sheath, or neurilemma*, while between the two is a quantity of white albuminoid matter, which undergoes coagulation, and is termed the *white substance of Schwann*. These fibres, with a few grey fibres, are united together to form those white bands commonly known as nerves. Some of them seem to commence in the tissues by free extremities, others by loops, many in cells which will be found most marked in the organs of special sense and in voluntary muscles. In the latter we find these cells under the form of expanded plates termed the *motorial plates;* the rods and cones of the eye, fibres of Corti of the ear, and Pacinian corpuscles of the skin, are cells of special sense. *Nerve-cells* are found in the nerve-centres, they are sometimes spheroidal, but frequently present one, two, or more prolongations from which the axis cylinders of the nerves pass. They are termed *polar cells* (*multi-, bi-*, according to whether they have many or but two prolongations). They abound in the grey portions of the centres, the white being mainly composed of nerve-fibres, connective tissue, and granular matter. Nerve-cells present each a nucleus with one or more nucleoli surrounded by granular matter sometimes of a dark colour.*

We have seen examples of **pure membrane or lemma** in the investing layers of the nerve and muscular elements termed respectively *neurilemma and sarcolemma*. A third form is brought under our notice in investigating the complex membranes of the body. It is termed *basement membrane*, and consists of a thin structureless or slightly granular fibre on one side in contact with connective tissue with numerous vessels, on the other with cells with more or less marked walls, nuclei, and nucleoli, termed epithelial

* When a number of nerves passing from different parts of a centre become mixed at one part, and from this nerves, each composed of fibres from several parts of the centre, run to distinct organs, we term the arrangement a plexus.

cells. The **compound membranes** are together made up therefore of internally a vasculo-areolar layer, centrally basement membrane, externally epithelial cells. They are of four kinds—serous, synovial, mucous, and cuticular.

Serous membranes are the most simple. They line closed sacs, do not secrete fluid from the blood, but merely are moistened by some halitus or vapour which serves to allow them to move freely upon each other without friction, for two layers of serous membrane generally come in contact whereby motion of the contents of a cavity without motion of its walls is admissible. The epithelium of this form of membrane consists of a single layer of cells spread over the surface of the basement membrane, closely fitted together, whereby each assumes the hexagonal form characteristic of *tessellated epithelium*. The pleura, pericardium, and arachnoid are good examples of serous membranes. It is supposed that they are merely dilated extremities of lymphatics. **Synovial membranes** differ from serous in their greater complexity and in the nature of their secretion, for they separate from the blood a fluid which, in consequence of the fatty matter it contains, serves to lubricate freely movable joints. They therefore line the cavities of such joints covering the inner surface of its ligaments, and in young animals even extending over the cartilage of the articular surfaces of the bones. In some parts, either to fill up space or to afford increased secreting surface, the vascular layer of the membrane is thickened, rendering it prominent and heightened in colour, forming the **synovial fringes**. Besides lining the cavities of joints synovial membranes form the **thecæ or sheaths of tendons**, lining the channels through which they pass, and being reflected over the tendons, the two portions being connected by double folds of membrane termed *frœna*. Whenever a tendon plays over a prominent surface whereby the direction of its action is altered it has placed between it and the prominence a small synovial closed sac termed a **bursa**, the third form of synovial cavity ; a marked bursa is situated between extensor pedis tendon and the antero-inferior part of metacarpi magnum, the synovial apparatus between flexor brachii and the humerus is intermediate in character between a theca and a bursa. Sometimes two or even three of the different forms of synovial cavity communicate. Synovial membranes always present simple *tessellated epithelium*. It is on the **Mucous**

membranes that we find it gaining its greatest complexity, in some cases consisting of several superimposed layers, the most internal of which, drawing nutriment from the blood, "hands it on" to those more externally placed. This is *stratified epithelium*. In some cases, however, as in the guttural pouch, we find the mucous membrane resembling a serous membrane in the simplicity of its component elements. The vascular layer of mucous membrane is termed *corium*, and is sometimes thick, presenting numerous and large blood-vessels, as in the case of the mucous membrane of the hard palate. In some situations it is extremely attenuated or even, as in the corneal conjuctiva, completely absent. Epithelium of mucous membrane is of two kinds, *ciliated or non-ciliated*. **Non-ciliated epithelium** may be either *tessellated* (above described) or, the mutual compression of the cells not being so great, they may retain the *spheroidal* figure (as in the bladder, ureters, and pelvis of the kidney), or the cells may be cylindrical, fitting in closely together, having one extremity free, the other in contact with the basement membrane. This is *columnar epithelium;* it lines the greater portion of the digestive tract. Either columnar or spheroidal cells may present on their free surface a number of small pointed prolongations which, by some peculiar power independent of nerve influence, wave to and fro. These are ciliæ, their presence is manifest in the respiratory tract, where their action serves to waft upwards mucus for expectoration ; and also in the Fallopian tube, where they waft the sperm-cell upwards, the germ-cell downwards. Mucous membrane lines all those great cavities of the body which communicate with the external air, but *the distinction between the mucous and serous membranes is very artificial*. Thus, the alimentary, respiratory, genito-urinary and lachrymal tracts are all lined by mucous membrane, and as this insensibly blends with the skin at the external openings of the body, we may look upon the structures of the body already described as contained in a large sac of compound membrane. In some parts the corium of mucous membranes presents unstriated muscular fibre, in others it is thickened at certain points producing *papillæ*, such as those found on the tongue and the intestinal mucous membrane. Other peculiarities in the mucous membranes will be noticed in the descriptions of different parts ; thus, the teeth, as will be shown, *are modified mucous membrane.* The **skin** consists of exactly the same component elements as other membranes, but they are generally more complex. It is in many parts continuous with the larger mucous membrane, and all over its surface presents openings leading into glands. It presents modifications under the form of hair and hoof. It consists of two parts separated by basement membrane, *dermis and epidermis.* The **dermis** corresponds to the corium of mucous membrane, and consists of a more or less thick layer of reticulated yellow elastic and white fibres, containing in their meshes vessels and nerves. It presents two layers, the *reticulated and papillated.* The **reticulated layer** internally situated, its deep-seated surface is continuous with the subcutaneous areolar tissue, its superficial with the **papillated layer**, which is externally raised into a number of finger-like prominences (covered with basement membrane), on to which the epidermis is moulded. The *subcutaneous areolar tissue* varies in its density in different parts of the body, being in some places very loose, as in the intermaxillary space, on the chest, &c. These parts are most liable to dropsical collections and most readily admit insertion of setons. It contains collections of fat, the terminations of the hair-follicles and of the sudoriparous glands. Portions of these also, and the sebaceous glands, are embedded in the dermis. The **epidermis** consists of epithelial cells moulded upon the external surface of the dermis; the most deep-seated are spheroidal and contain pigment-granules, forming the *rete mucosum.* The superficial layers become hard, dry, and scaly; as they are subjected to external influences, they are gradually worn away.

Hairs are produced from depressions on the surface of the skin termed **hair-follicles.** These are filled with spheroidal epithelium, and into each of them open generally the ducts of two sebaceous glands. At its deep-seated extremity, where it forms a cul-de-sac, each follicle presents a papilla from which the hair is produced. The hair is the epithelium of the papilla (which is of the same nature as the dermis). It consists of a *bulb, apex, and shaft.* The *bulb* immediately surmounts the papilla, and is the largest portion of the hair, consisting of the softest and most recently formed spheroidal cells. The *shaft* is the cylindrical portion which runs outwards, emerges from the follicle, and terminates in the *apex*. It consists of an external or *cortical layer*, the cells of which are imbricated (arranged like tiles upon a house), and a central *medullary portion*, consisting of soft, spheroidal cells. The **sebaceous glands** pour their secretion into the follicle; it is an oily substance, and permeates the hair, rendering it pliable. These glands are small flask-shaped bodies of the racemose class. The **sudoriparous glands** produce the *sudor or sweat*, which they pour out upon the surface of the skin. Each gland consists of a single elongated tube, which is very much twisted upon itself in the subcutaneous areolar tissue so as to form a small, rounded mass permeated by blood-vessels. It takes a slightly wavy course through the dermis, and in passing through the epidermis becomes cork-screw like. Hairs are either short and glossy, as those covering the surface of the body in general, or long and coarse, with their bulbs deeply imbedded in the subcutaneous tissue, forming *horsehair*, found in the mane, tail, eyebrows, eyelashes, and tentaculæ of the lips, which in the upper lip of some coarse-bred horses become long and wavy like the human mustachois. Fineness and scantiness of hair is a mark of a well-bred horse. It varies in colour in horses from white to black. In the ass it generally assumes a brownish-white mouse colour, and a dark line of hair extending down the back assists in distinguishing Equus asinus from E. caballus. The modification which hair undergoes in forming horn will be described in its proper place.

We must now examine those complex arrangements of mucous membrane which produce **glands**. All glands are depressions in mucous membrane, and these depressions may assume either the *tubular or the follicular form*, and either of these may be *simple or compound*. **Simple tubular glands** are those which consist of a single undivided tubular depression of mucous membrane: examples are the follicles of Lieberkühn, and the sudoriparous glands—the former of which are extremely shallow, the latter much elongated. Sometimes these tubes divide at their deep-seated extremities, and each part may again subdivide, and thus, by continuation of this process, **compound tubular glands**, such as the testes or the kidneys, are produced; the latter organ has dilated extremities of its tubules, and thus resembles compound follicular glands.

Simple follicular glands are flask-shaped bodies opening on to the surface by a small duct; when several of these pour their secretion into a compound duct, we see the first example of a **compound follicular or racemose gland**, which consists of a number of flask-shaped or vesicular bodies, which pour their secretions through a number of ducts, which unite and reunite until at last they form a more or less elongated canal, the *duct of the gland*, which opens on to the surface of the mucous membrane lining the cavity to which the gland is appended. The *secreting vesicles or acini* thus bear to the ducts the relation which grapes bear to their stalks, hence the name racemose. The several parts of the gland are closely invested by areolar tissue, whereby they are united to form *lobules*, which by their union complete the gland. As examples of compound follicular glands, may be adduced the salivary, pancreas, lachrymal, and mammary glands, while the meibomian, brunners, and the sebaceous glands present this type under a much less complex form.

The so-called **ductless glands** present peculiar characters, which will be specially noticed. All the above-mentioned structures are supplied with matter for growth and for repair by the blood, and the same fluid also removes matters for which they have no longer use. It is conveyed from the heart throughout the system by certain vessels found in all parts of the body more or less. It must, however, be clearly understood that in all cases the special cells of a tissue are externally placed to the blood-vessels, so that all nutriment obtained by any but blood-cells must be obtained through the walls of the vessels and through the wall of the cell. The vascularity of a part consists in the close relation of its cells to its vessels. Thus the capillaries run between the muscular fibrillæ of voluntary muscle; they are separated from the cells by basement membrane in the complex membranes; in stratified epithelium the outer layers are separated by basement membrane and a number of cells which hand on the nutritive portion or plasma of the blood from within outwards. The same condition obtains in cartilage except that the cells are separated by intercellular substance of variable nature and amount; while in bone the plasma passes through an elongated but extremely minute canal in its course from the vessel to the cell. The vessels from which the cells directly derive the plasma are the **capillaries**, extremely minute vessels, uniform in size, forming an intricate network, varying in the figure of its interspaces in the different tissues, which in muscle are elongated parallelograms, in lungs smaller than the vessels themselves. These vessels connect the terminal extremities of the arteries with the commencement twigs of the veins ; the blood does not take any definite course through them. The wall of the capillary consists of simple membrane with more or less numerous nuclei imbedded in its substance. When the vessels are large they present nuclei elongated both longitudinally and transversely ; the smaller vessels have only longitudinal nuclei, which are widely scattered. The brain and the mucous membrane of the bowels present extremely minute capillaries, those of the skin are very large (as may be well seen in that part modified to form the hoof). In the spleen, corpus cavernosum penis, and in the chorion, the arteries open directly into veins usually modified to form venous sinuses.

The **arteries** carry blood from the heart to the tissues. They vary in size in proportion to their distance from the heart. They may be distinguished from the veins by their elasticity, which enables them to remain open after death, and also to expel the blood which they contained when the heart's action ceased, whence they are found empty. The sum of the areas of the arteries at a distance from the heart is greater than the area of the aorta, and the size of this sum is great in proportion to the distance from the heart, so that the arterial system has fancifully been described as a cone, the apex of which is at the aorta, the base at the capillaries, in which the smallest arteries terminate. Arteries may be given off at right angles or at an acute angle from a large trunk; as the pressure of fluids in closed vessels is equal in all directions, this arrangement will scarcely influence the flow of blood to particular parts. Some arteries run directly to the organs which they supply, others, especially the spermatic artery in the horse, assume a peculiar convoluted character ; while in some cases several vessels unite to form a more or less intricate plexus termed *rete mirabile*, examples of which may be seen in the circle of Willis and the circular arteriosus of the foot. Arteries possess *three coats*. The *external or fibro-areolar coat* consists of more or less condensed areolar tissue, with a slight predominance of elastic fibres; it serves to protect the vessels and to form the nidus in which the small nutrient vessels of the arterial coats (*vasa vasorum*) break up into capillaries. It is directly continuous with the areolar tissue of the channel through which the artery runs, which sometimes becomes developed into a marked sheath enclosing other structures in addition to the artery; the carotid sheath is an example. The *middle or musculo-elastic coat* consists of unstriated muscular fibre in union with yellow elastic tissue. The muscular fibres assume a direction transverse to the longitudinal axis of the vessel; they predominate in the smallest arteries, those most distant from the heart. The elastic fibres mainly take a longitudinal direction, but are arranged in the form of a network; in the largest arteries, as the aorta, they form the major portion of the arterial wall. The *internal coat* is of a serous nature, presenting an internal layer of tessellated epithelium and *a peculiarly modified basement membrane*. The latter, termed *the fenestrated membrane of Henle*, shows a peculiar tendency to curl up when removed from its situation, and it has numerous perforations. These characters seem to indicate its intermediate nature between common basement membrane and ordinary yellow elastic tissue with which in some vessels its external surface intimately blends. This coat of the artery is continuous centrally with the endocardium, peripherally with the constituent membrane of the capillaries. The course of arteries is in some cases irregular, either as a result of disease or of original formation. Descriptions of these cannot be accurate *for all cases*. When arteries run near one another they almost always *anastomose or inosculate*, communicate more or less closely, either directly or by branches. Hence provision is made for continuance of supply of blood after obliteration of one source; the collateral sources after injury resulting in obliteration of an important artery generally prove equal to the emergency—a point of great value in surgery.

The **veins** carry blood from the capillaries to the heart with exception of the portal veins, which convey it from the alimentary organs of the abdomen to the liver, where it is purified prior to admission into the general round of the circulation. Arteries inosculate or anastomose in some situations freely, but veins rarely approach each other without communication; they are also more numerous, much less regular, and larger than arteries. A portion of vein removed from the body will be found more or less coloured by blood which has remained in it after death. The thinness of its walls will allow it to collapse when placed upon a flat surface. Like arteries veins have three coats, which differ from those of the arteries in the paucity of their component elements. The *internal or seroid coat*, however, requires notice, for, in addition to lining the vessel, it projects into the cavity, double folds constituting *valves* of a semilunar form. These always have their free margins and concave surfaces directed towards the heart, for they perform an important function in promoting the circulation. According to the size of the vessel from one to three valves may coexist at one particular part. Only two arteries, aorta and pulmonary artery, possess valves, and they look from the heart. Some veins, as those of the foot, portal system, and several components of the anterior and posterior venæ cavæ, have no valves; in other situations the valves are few in number. When not in use during the forcible rush of blood through the vessels, the valves lie against the sides of the vessel. This is their ordinary situation in the jugulars, in which they come into action only when the animal's head is lowered, as in grazing; under other circumstances the blood passes to the heart in virtue of its own weight. Certain modifications of veins, termed *venous sinuses*, will be noticed in due course.

In addition to blood-vessels we find other vessels which convey a fluid from which blood is prepared. The free spaces of the body, and, according to some authorities, even the serous cavities, are lined by an epithelial membrane, continuous with that lining certain vessels termed **lymphatics**, which unite and reunite until at length they pour their contents into the blood-vessels. These vessels probably run from all parts of the body; the largest trunks which they form are the *thoracic duct* and the *great right lymphatic*, which open into the large veins at the anterior part of the chest. The smallest lymphatics resemble the capillaries in structure, the largest closely resemble veins, presenting valves, and passing towards the centre of circulation, but they anastomose less freely, their valves are less numerous, the fluid they contain is colourless, and sooner or later almost all of them pass through certain dark grey nodular bodies, found in various parts of the body, termed lymphatic glands. These produce a change in the nature of the contained fluid, and seem to be mere means of economy of space, since some of the lower animals in place of glands present extreme elongation and tortuosity of the lymph vessels. In these lower animals the lymph vessels present in some parts dilatations with thick muscular walls, **lymph hearts**, which drive the lymph into the blood-vascular system *in several situations*. In the higher vertebrata the communication between the blood and lymph systems seems to occur only at the anterior part of the thorax. Some authorities have described these large lymph vessels as contractile.

Lymphatic glands consist of a special investing coat with trabeculæ passing inwards, dividing the contained space into cavities, of which the external are most marked, all of which are lined with epithelium. The external portion, therefore, is softest, and is termed the *cortical portion*; the firmer central structure is the *medullary structure*. A number of beaded *afferent lymphatics* converge and pass into a gland; they seem primarily to pass to the medullary portion in which they are arranged in a convoluted mass. They are supposed to open into the cavities of the gland, *lymph spaces*, and the lymph to pass irregularly through these until it arrives at the commencement portion of the *efferent lymphatics*, which, larger and less numerous than the afferent vessels, on emerging from the gland pass towards the main lymphatic trunks.

The lymph vessels of the bowels are distinctively called the **lacteals**, for the fluid they convey is rendered opaque and milky in appearance by the presence of fatty globules taken up from the contents of the intestines by these vessels. They differ in no respect from other lymphatics; they commence in the villi of the intestines either in blind extremities, in loops, or in plexuses.

PART II—OSTEOLOGY.

THE SKULL,

with a few cartilaginous accessory structures, forms the basement structure of the head. It is composed of modified vertebræ, united together in such a manner as to form numerous cavities, mostly for the purpose of protecting important organs. It is a conical body, the apex part or **muzzle**, of which we shall consider as being directed forwards, the base towards the trunk to which it is united by the neck. It is divided into the **cranium**, which corresponds to the neural portion of the other vertebræ, and the **face**, which is antero-inferiorly placed, and corresponds to the hæmal ring. The latter much predominates in size.

THE CRANIUM

consists of twelve bones, four pairs and four single. The cavity formed by these bones will be considered when its contents are to be examined; we shall here examine as distinct bones those osseous masses which, in the adult, are firmly united into a single mass with those of the face forming the **upper jaw**. The **lower jaw** is formed by a special bone of the face (the inferior maxillary bone). The union between the upper and lower jaws, and that between the cranium and the first cervical vertebra, are the only examples of synovial articulations to be found in the skull. The union of os hyoides (the tongue bone) with the base of the skull is fibro-cartilaginous. All the other joints are immoveable, and sooner or later become obliterated.

OS OCCIPITIS is the postero-superior bone of the cranium, that which articulates with the atlas, and which forms the "poll." It is divided into *body, condyles, basilar process, and styloid processes.* It is a single bone lying along the middle line of the body, and its **basilar process** forms the posterior part of the base of the cranium. It is elongated from behind, where it blends with the condyles, forwards, where it is roughened, coming in apposition with os sphenoides. It widens posteriorly in extending upwards towards the roots of the styloid processes, externally to the condyles, and in these parts we see a large foramen on each side, *condyloid*, through which the lingual nerve passes. We shall simply in future note the position of the foramina of the skull, as a more detailed description may be found elsewhere.

The *under surface* is convex, and anteriorly presents roughened depressions for attachment of rectus capitis anticus minor. Posteriorly it presents a small smooth process, extending between the condyles, expanding posteriorly to blend with the upper surface in forming the inferior part of foramen magnum. The *upper part* is concave in all parts, most so posteriorly, where laterally it presents the upper openings of the condyloid foramina; while anteriorly its external margins are sharpened, each forming the inner boundary of *foramen lacerum basis cranii*. The **condyles** are two in number, forming the lateral boundaries of foramen magnum. They are articulatory prominences, convex in all parts, superiorly looking upwards and slightly outwards, inferiorly downwards and outwards. Their inner aspect is upwards and inwards; superiorly they join the body of the bone. From their external angles run the styloid processes; inferiorly they are continued on to the basilar processes, converging, but separated by a smooth concave surface. They are mainly separated from the styloid processes by deep fossæ, which present the condyloid foramina. The **styloid processes** are long, and extend in a downward direction, being slightly curled inwards, and flattened from without inwards. They afford attachment to stylo-hyoideus, stylo-maxillaris, digastricus, and obliqui capitis superior and anticus, and one of the attoido-occipital ligaments. Their anterior surface at the superior part is roughened for articulation with os temporale petrosum, and superiorly their margin is continued upwards as the *mastoid ridge* to the true crest of the bone. The **body** of the occiput is continued upwards from the superior part of the condyles and styloid processes laterally, centrally from the superior margin of foramen magnum, which it forms by means of a concave, somewhat sharp margin. It presents three surfaces: the *anterior internal surface* is firm, and presents a number of depressions for accommodation of the cerebellar convolutions, from which it is separated by the membranes of the brain. The *posterior surface* is bound on either side by the mastoid ridges, and meets the superior in forming a prominent ridge with a backward inclination, **crest or crista.** This surface is divided into two parts by a roughened line extending from the centre of the crest towards foramen magnum. To this the cordiform portion of ligamentum nuchæ is attached; the concave roughened surfaces on either side of it, **scabrous pits**, afford attachment to complexi major and minor and recti capitis major and minor postici. To the crest and mastoid ridges are attached temporalis, obliquus capitis superior, levator humeri, trachelo mastoideus, splenius, and (to the crest) some of the auricular muscles. The superior surface of the bone forms the posterior part of the forehead, and is roughened, for from it commence the parietal ridges. Thus it affords attachment slightly to temporalis on either side, and anteriorly presents an irregular margin dentated centrally for union with os triquatrum, squamous laterally for union with the parietal and more externally with the squamous temporal bone.

OS TRIQUATRUM is a small shield-shaped portion of bone, with a smooth and flat superior surface and dentated margins; *antero-laterally* convex, meeting at a rounded point anteriorly for union with the parietals. *Posteriorly* concave, thus presenting two prominent posterior angles, articulating here with os occipitis. Its *inner surface* presents the **ossific tentorium**, a remarkable bony prominence, with three concave surfaces, uniting to form three sharp margins, one at each side, giving attachment to the membranous tentorium, one anteriorly for attachment of the posterior margin of falx cerebri.

The **PARIETAL BONES** are flat and thin, presenting two surfaces and five margins. The *external surface* is convex, looking upwards and outwards. At its posterior part it presents a few foramina for the passage of meningeal arteries to the coverings of the brain. It is divided into two parts by the **parietal ridge**, a slightly prominent line of bone passing obliquely outwards from the anterior part of the crest of the occiput towards the orbital process of the frontal bone, and which gives attachment to auricular muscles. The outer division helps to form the temporal fossa, and affords attachment to temporalis; it is continued by a thin, very rough squamous surface, on which the squamous temporal bone rests. The inner portion of the external surface is immediately subcutaneous. The *postero-internal margin* articulates by suture with os triquatrum, the *inner margin* similarly with its fellow. The *anterior margin* inwardly presents a rough surface for squamous articulation with os frontis, outwardly the outer layer of the bone projects, so that it rests on os frontis. The *external margin* forms the thin jagged edge of the squamous suture with the squamous temporal bone, as also does the *postero-external margin*, which meets with the postero-internal, forming a point which articulates with os occipitis. The *inner surface* of the bone is mostly covered with digital impressions for accommodation of the cerebral convolutions, but along the postero-external margin has a smooth groove through which a large vein runs to torcular herophili, from a small foramen just above the mastoid process of squamous temporal bone. Below this groove, at its outer part, is a rough surface for union with the petrous temporal bone.

OS FRONTIS (one of which is found on each side of the cranium) forms part of the anterior and of the superior portion of the cranium. It also forms the supero-posterior part of the nasal chamber and a considerable portion of the inner wall of the orbit. So it presents a cranial, a facial, and an orbital plate. The *facial plate* is quadrilateral, smooth externally, internally at its posterior part, blends with the cranial plate, at its anterior is roughened, helping to form the frontal sinus. The outer surface is flat, in parts convex, and subcutaneous with exception of its *outer margin* which is covered by orbicularis palpebrarum. From the posterior part of this margin projects the **orbital process**, a long prominence, flattened from above downwards, externally convex in contact with the orbicularis palpebrarum, internally concave in contact with the lachrymal gland, having periorbitale attached to its posterior margin. Its outer extremity articulates with the zygomatic process of the squamous temporal bone; its anterior margin forms the superior part of the rim of the orbit, its root becomes expanded in becoming attached to the outer edge of the facial plate, and presents the *supra-orbital foramen*. The anterior part of this edge forms the inner part of the rim of the orbit, and is continuous downwards with the orbital plate. The *anterior margin* presents a laminated squamous surface to the outer part of which the lachrymal, to the inner the nasal bone becomes attached. The *inner margin* presents a roughened edge for apposition with its fellow, from the anterior part of which a thin plate of bone is reflected downwards to form the inner boundary of the frontal sinus lying in contact with its fellow. The *posterior boundary* is squamous, most so on the outer side, for articulation with the parietal bone. The **cranial plate** blends superiorly with the posterior part of the under surface of the facial plate, and presents digital depression for the cerebral convolutions. Its *inner margin* is serrated and lies in contact with its fellow. Its *inferior margin* joins the upper part of the crista galli and cribriform plates of the ethmoid. Its *outer margin* joins the orbital plate. The *anterior surface* helps to form the frontal sinus, and from the *inner margin* of the surface runs the vertical bony septum which inferiorly is reflected to form a narrow ledge of bone running obliquely forwards and upwards in blending anteriorly with the facial plate in a sharp point. The **orbital plate** externally is concave, forming the inner part of the orbit. Near the root of the orbital process it presents the fibro-cartilaginous loop through which the superior oblique muscle of the eye plays. Posteriorly it is continued below the orbital process, helping to form the temporal fossa, and articulating with the squamous temporal bone. This plate presents a deep indentation, inferiorly extending upwards towards the orbital loop. It terminates in a point, and accommodates the orbital surface of os ethmoides, which, by the older anatomists, was termed *os planum*. The *anterior margin* inferiorly gives attachment to os palati; centrally to the superior maxillary bone; superiorly to the lachrymal bone. The *inner surface* of this plate posteriorly is separated by a remarkable groove from the cranial surface, and has a squamous suture with the ala of the ethmoid; centrally it gives attachment to the cranial plate; anteriorly forms the convex outer wall of the frontal sinus. At the extreme postero-inferior part this plate in articulating with os ethmoides forms foramen orbitale internum.

OS ETHMOIDES forms a considerable part of the anterior boundary of the cranium, and consists mainly of a *body*, from which an *ala* projects on either side, and a remarkable process projects upwards anteriorly. The **body** articulates posteriorly with the body of os sphenoides: here it is rounded and solid, but anteriorly it presents two large cavities, **ethmoidal sinuses**, indirectly continuous with the nasal chamber. The *under surface* presents centrally a small convex smooth surface, on either side of which is roughened for squamous suture with the vomer. Each of the *sides of the body* anteriorly is occupied with a squamous surface for os palati, posteriorly by a continuation of that on the posterior part for os sphenoides. The *upper surface* looks in an upward and slightly backward direction, and is concave. Posteriorly it presents a peculiar depression separated from the rest of this surface by a sharp ridge, **optic fossa**, from which the two foramina optici pass through the substance of the bone, for passage of the optic nerves, running from the optic decussation, which rests in the fossa. The **wings** look inward and slightly upwards. Their *upper surfaces* are completely occupied in formation of the cerebral cavity; they therefore present digital impressions. At the line of junction of the *under surface* with the body of the bone posteriorly is the opening (inferior) of the optic foramen, and in front of this is a smooth groove, which by union with a similar depression in os sphenoides forms foramen lacerum orbitale. The *outer margin* of this surface of the ala is mainly occupied by a squamous suture for union with the special groove in os frontis; and the smooth space between, by some anatomists described as a distinct bone, *os planum*, serves to form a considerable part of the inner wall of the orbit, fitting into the deep depression extending from the inferior margin of the orbital plate of os frontis. From the antero-superior part of the body **crista galli process** runs in an upward direction, to join superiorly the infero-central part of the cranial plates of ossa frontis. Its *superior extremity* is enlarged, and presents a concave triangular surface for os frontis. Its *posterior margin* is sharp, smooth on either side and affords attachment to the anterior extremity of falx cerebri. *Anteriorly* it presents a more or less prominent ridge, extending from above downwards, continuous anteriorly as septum nasi. On either side it is jagged, and continues as far as the anterior margin of the wings by the **cribriform plates**, perforated thin layers of bone through which the fibres of the olfactory nerves pass, and which *posteriorly* form the boundaries of depressions **olfactory fossa**, in which the olfactory bulbs rest; *anteriorly* have attached to them certain thin plates of bone curled upon one another in a complex manner, forming the **ethmoid cells**, a direct continuation of the ethmoidal sinuses in the body. These are portions of bone which derive their nutriment directly from the mucous membrane which covers them, having no special periosteum. In that mucous membrane (and also that of the superior turbinated bone) the fibres of the olfactory nerve are distributed, rendering it the organ of the special sense of smell.

OS SPHENOIDES forms the major portion of the floor of the cranium, and is situated between the occiput and os ethmoides. It consists of *a body and two wings with two crura or legs*. The *upper surface* of the **body** is irregularly concave, receiving the name **sella turcica**. Centrally it supports the pituitary body, and on either side of this is a depression in which rests the cavernous sinus, while extending on to the upper surface of each wing on each side we see a groove for the superior maxillary division of the fifth cranial nerve, running forwards to join the *subsphenoid foramen in forming foramen rotundum*. This is separated anteriorly by a thin plate of bone from a smooth groove, which unites with another on os ethmoides to form *foramen lacerum orbitale*. The *posterior extremity* presents a roughened ovoid surface for articulation with the anterior extremity of the basilar process of the occiput. *Anteriorly* is a similar surface for apposition with the posterior part of the body of the ethmoid, which is separated on each side from foramen rotundum by a thin plate of bone, which presents a small squamous surface for the antero-internal part of the wing of the ethmoid. The under surface of the body is convex, tending to become smooth anteriorly. Posteriorly it is rough for insertion of a rectus capitis anticus major on either side. The wing of os sphenoides superiorly is concave and presents an irregular surface, helping to form the cerebral cavity. Its posterior margin is internally sharp, externally rounded, forming the anterior boundary of foramen lacerum basis cranii. Its *superior margin* presents externally a squamous serrated surface for the squamous temporal bone. In the angle which it forms with the anterior margin is foramen pathetici. The *anterior margin*, on its inner edge, articulates with the wing of os ethmoides, superiorly presenting the smooth groove which forms foramen lacerum orbitale, more externally foramen rotundum, and on its outer edge a ridge to which periorbitale is attached, which terminates inferiorly in an elongated process, flattened from side to side. This is the **crus** of the bone. Its *superior margin* and *anterior extremity* articulate with os palati ; its *inner surface* posteriorly with os pterygoideum, and to its *inferior margin* are attached masseter internus and pterygoideus.

OS TEMPORALE SQUAMOSUM (one on each side of the head) is situated at the postero-lateral part of the head, and is so named because its main portion is almost wholly occupied in squamosal connection with the neighbouring bones. It is divided into two parts—body and zygoma. The *inner surface* of the **body** superiorly along its whole length articulates with the parietal bone. Its *anterior margin* joins os frontis ; antero-inferiorly it joins the wing of os sphenoides ; postero-inferiorly it presents a deep notch, into which the petrous temporal bone with the external auditory process fits. The *posterior margin* unites with the petrous temporal bone, and at its extreme posterior part with os occipitis. The central part of the inner surface is smooth, and assists to form the wall of the cranium. The *external surface* is divided into two almost equal parts by a line running from the postero-superior to the antero-inferior angle, from the anterior part of which the zygoma commences. The space above this line helps to form the temporal fossa, having temporalis attached to it, while superiorly it presents some foramina. The inferior division of this surface is smooth and centrally presents a groove, the inferior opening of the canal between this and the parietal bone. Just in front of this is a small rounded prominence, the **mastoid process,** to which is attached one of the ligaments of the temporo-maxillary articulation. The **zygoma** is a large process, which is very thin, flattened from above downwards, running directly outwards at its origin. It is then twisted from below upwards (so that its outer margin becomes superiorly placed), and runs in a forward direction. Its *superior margin* anteriorly presents a roughened prominence, to which the outer extremity of the orbital process of os frontis is united. In front of this, the *inner surface*, is concave, forming part of the orbit; the *outer surface* presents a squamous suture, terminating anteriorly in a point for the zygomatic or malar bone. The rest of the outer surface affords attachment to masseter externus. The *superior surface*, at its junction with the body, forms a deep, smooth, rounded groove, in which the coracoid process of the inferior maxilla fits. The *under surface* presents the glenoid cavity coated with articular cartilage, extending from without inwards on to the body, rounded posteriorly by the mastoid process; it articulates with the condyle of the inferior maxilla through the medium of a disc of cartilage.

OS TEMPORAL PETROSUM (Pair) is an extremely irregular bone, wedged in between the occipital, parietal, and squamous temporal bones. Its *posterior surface* lies in contact with os occipitis, its *anterior* superior with the parietal, inferiorly with the squamous temporal bone. It is the hardest bone in the body, and *contains the organ of hearing* (as will be afterwards described). Its *inner surface* helps to form the inner surface of the cranium, and presents digital impressions, with centrally *foramen auditorium internum*. The *external surface* is very rough and irregular; superiorly it assists in forming the mastoid ridge, giving attachments to several of the cervical muscles; centrally it presents the **external auditory process**, to which the annular cartilage of the ear is attached, and which is perforated by the *external auditory hiatus*, which extends inwards as far as the tympanum. Below and in front of this there is a prominence, which looks as if a round peg had been driven into the bone; this is the **hyoid process**, to which the antero-superior angle of the long cornu of os hyoides is united by fibro-cartilage. Immediately below this is a rounded, roughened protuberance, denoting the situation of the *mastoid cells* inside the bone; hence it is termed the *mastoid process*. In front of it is an elongated, sharp process, the **styloid process**, to which tensor palati and stylo-pharyngeus are attached, and just below and behind it is the *styloid foramen*, from which the Eustachian tube passes to the pharynx.

BONES OF THE FACE

are nine pairs and two single bones. We shall first examine the pairs.

OS NASI is a flat bone, forming the major part of the roof of the nasal chamber; presents two surfaces, two margins, a base, and an apex. The *base* of this bone has a squamous laminated surface for union with os frontis. The *inner margin* is straight, and lies in apposition with the corresponding part of its fellow. The *outer margin* meets it at an extremely acute angle, and as the anterior part of the outer margin is free, the two points coming together produce a remarkable prominence, to which the nasal cartilages are attached, and which is termed the **nasal peak**. From the *external surface* of the bone, near the outer margin of the nasal peak, dilator naris superior arises; behind this the margin is in apposition with the upper extremity of the superior process of the anterior maxillary bone. It then presents thin horizontal plates fitting into os maxillare superius, and at the outer angle of the base is a line of union with os lachrymale. The superior surface is convex; inclined obliquely outwards over it passes nasalis longus labii superioris. The *inner surface*, correspondingly concave, is partly occupied by the expanded upper portion of septum nasi, but inclined to its outer side it presents a *ridge to which is attached* os **turbinatum superius**. The turbinated bones are extremely thin and friable in the dried subject, but when fresh they are coated on both sides by mucous membrane, which renders them elastic. Each of them consists of two portions, one anteriorly placed, belonging to the nasal chamber, the other to the sinuses of the head. These bones are supplied with nutriment by their investing mucous membrane; they therefore have no periosteum. They will be noticed more particularly in our description of the nasal chambers and the sinuses of the head.

Os turbinatum inferius is attached to the nasal surface of the facial plate of the superior maxillary bone.

Os maxillare anterius is situated at the anterior part of the upper jaw, lodges the superior incisors, and forms the outer and inferior boundary of the anterior opening into the nasal fossæ of the skeleton; it consists of a body and two processes. The *body* is externally convex, affording attachment to levator labii superioris. Its *internal surface* lies in contact with its fellow by a broad roughened surface, and the two bones here together form *foramen incisiorum*, running from the roof of the mouth at its extreme anterior part. The anterior attachment of septum nasi extends on the bone as far as this foramen. The *inferior margin* of the body presents on each side three **alveolar cavities**, into which the incisors fit. These are conical depressions, separated from each other by thin bony plates; thus the two central incisors have between them two bony laminæ. Behind the convex alveolar line is a smooth portion of the lower margin of the bone, and behind this is a *portion of the alveolar cavity for the tush*, which is completed by the superior maxillary bone. From the supero-posterior part of the body an **elongated process** passes obliquely upwards and backwards. Its *superior margin* runs from the upper portion of the symphysis, first outwards, then upwards, as far as os naris. It is rounded, to its external surface dilator naris inferior is attached, and it winds round towards the false nostril. Its *posterior margin*, inferiorly by squamous suture, superiorly by schindylesis, is in connection with the superior maxillary bone. The *upper margin* is united to the anterior part of the roughed portion of the outer margin of the os nasi. The *inner surface* forms the anterior margin of the lateral boundary of the nasal chamber. The *inferior surface* of the body of this bone slopes obliquely upwards and backwards, and centrally terminates in a thin process, the *palatine*, running backwards towards the palatine process of the superior maxillary bone. It is flattened from above downwards; its extreme postero-superior parts in some cases is in apposition with the anterior extremity of the vomer; it forms the inner boundary of a space, *incisive opening*, which is generally filled with cartilaginous matter continuous superiorly with septum nasi, embedded in which is the organ of Jacobson.

Os maxillare superius is a large bone, with its fellow forming a very considerable portion of the face. It assists in forming several cavities, and completely forms one of the facial sinuses, the inferior maxillary. These sinuses, however, are only found markedly developed in the adult. It presents two surfaces and four margins. The *superior margin* is irregular, being in apposition with the outer margin of the os nasi; it is but half the length of the inferior. The *anterior margin* slopes obliquely downwards and forwards, and joins the posterior margin of the nasal process of the anterior maxilla. It meets the *inferior margin* at an acute angle in forming the posterior boundary of the alveolar cavity for the tush. The *posterior margin* presents an irregular concavity looking upwards. It is a rough squamosal surface extending into the outer part of the bone, increasing in width from above downwards, superiorly in apposition with the lachrymal, inferiorly with the malar bone. This surface terminates above in a projecting point, which slightly meets the squamous temporal bone. The inferior surface of the projection is smooth for the passage of the superior varicose vein of the face; below, and slightly behind, is an irregular prominence, **tuberosity**, to which buccinator and retractor labii inferiosis are attached. The *inferior margin* along the greater part of its length is occupied by the alveolar cavities for the upper molars. These consist of a series of deep square cavities, separating the bone into two plates, and separated from one another by thin bony septa. Their walls correspond to the external surface of the teeth, into the grooves of which the bony matter is moulded. They are six in number; the first and last cavities of the series are triangular, and sometimes at the anterior angle of the first is a small alveolar cavity in which the premolar or "wolf's tooth" is situated. In front of this the inferior margin is smooth and rounded to its anterior extremity. The *external surface* slopes obliquely downwards and outwards. From about the centre of the articulatory surface of the posterior margin (which encroaches considerably upon the surface) the anterior extremity of the **zygomatic ridge** runs forward for a short distance. It terminates about opposite the second molar tooth; to its under surface are attached masseter externus and zygomaticus, just in front of its anterior extremity retractor labii superioris arises. Buccinator posteriorly and caninus anteriorly arise from this surface just above the inferior margin; and between the former muscle and the zygomatic spine is the superior varicose vein of the face. The *superior maxillary or infra-orbitar foramen* is situated above and in front of the zygomatic spine; through it the major portion of the dental branch of the superior maxillary division of the fifth nerve (after passing through the canalis infra orbitale, a special channel of the bone which runs through the maxillary sinus) emerges with an accompanying artery, both run forwards concealed from view by levator labii superioris alæque nasi. From the point of junction of the malar, lachrymal, and superior maxillary bones, nasalis longus labii superioris arises. A considerable portion of the *inner surface* of this bone is occupied in forming the outer wall of the nasal chamber. Along the central line extending from before backwards is a ridge to which the inferior turbinated bone is attached. At about the centre of the bone this terminates in bifurcating, forming two lines, one of which runs upwards and backwards, the other downwards and backwards, mapping out that portion of bone which forms the inferior maxillary sinus. The extreme inferior margin affords attachment to the gums of the roof of the mouth; higher up is a groove running forwards for the palatine artery, and above this a horizontal process, **palatine process**, projects inwards. It extends neither as far as the anterior nor as the posterior extremity of the bone, but anteriorly, gradually diminishes in size to terminate imperceptibly; posteriorly presents a squamous surface extending on to the small part of the tuberosity for union with the palate bone. This surface presents a smooth groove, which helps to form the *palatine canal*. The *free margin* presents numerous spiculæ of bone, which serve to keep it closely in apposition with the palatine process of its fellow, and the two together form a groove into which the anterior part of the inferior margin of the vomer fits. The *upper surface* of this process helps to form the floor of the nasal chamber, the *inferior surface* is covered by the mucous membrane of the hard palate, with its rich venous plexus. This bone assists in forming the superior maxillary sinus through which two bony canals pass, the canalis infra orbitale, and the bony covering of the ductus ad nasum.

OS LACHRYMALE forms the antero-internal part of the orbit, assisting in forming its margin, thus being bent upon itself in such a manner as to present two plates, separated by the concave marginal border, which is for the most part smooth, but centrally rather rough. The *external surface*, therefore, as a whole, is convex, but it consists of two concave plates. The *anterior* is the *facial plate*. It presents at about its centre a *tubercle* (lachrymal) to which the orbicularis palpebrarum is attached. Above this it gives partial attachment to labii superioris alæque nasi, below to nasalis longus labii superioris. The *posterior or orbital plate* is funnel-shaped, converging from all points to towards the lachrymal foramen, situated just behind the centre of the marginal border. To this bone near this foramen is attached the inferior oblique muscle of the eyeball, and in it rests the lachrymal sac from which the ductus ad nasum is continued forwards through a bony canal, running along and attached to the internal surface of the facial plate of the bone. The *inner surface* of the bone assists in forming the superior maxillary sinus, and is therefore covered with mucous membrane. The *anterior margin* is in contact with os nasi and also with the superior maxillary bone, which extends along the whole length of the *inferior margin*, along the inner edge, while the malar bone has a squamous union with a considerable portion of the outer edge. The extreme posterior part of this margin is smooth. The *superior margin* has a squamous union with os frontis, but at the posterior angle, where this margin and the inferior meet, the lachrymal bone comes in contact with os palati.

OS PALATI forms the anterior, posterior, and outer boundary of the posterior naris. It is very irregular, but mainly consists of a bony plate elongated from behind forwards. This plate is convex on its internal surface, and posteriorly is divided into two equal parts by a prominent ridge. The upper part is articulatory supero-posteriorly for union with the lateral part of the body of ethmoides; in front of this os frontis occupies a small space; os lachrymale, a still smaller in front of that; but the greater part of this division is covered with mucous membrane, assisting to form the ethmoid sinus. Below the articulation with os frontis is *foramen spheno-palatinum*, which is continued by a groove on to the inferior part, which, with the inferior part of the ridge, posteriorly is united to the pterygoid and (more inwardly) the vomer. Anteriorly this part becomes much prolonged and is smooth and concave, forming the posterior boundary of the posterior nares. It becomes twisted from below upwards at its anterior extremity, thus forming a peculiar process, **palatine**, the inner margin of which is smooth, rounded, concave, and coated with mucous membrane forming the anterior and external boundary of the posterior naris in the skeleton. Its free extremity unites with its fellow anteriorly at a symphysis, and thus the two bones form a "*palatine arch,*" to which, in the fresh subject, the fibrous layer of the fixed portion of the soft palate is attached. The free margin of the ridge on the inner surface of the bone is united to the vomer, and by the groove of that bone separated from its fellow. The external margin of the palatine process is convex, and presents a serrated edge for union with the palatine process of the superior maxilla. Posteriorly this roughened surface increases in size, presenting a very wide squamous surface for union with the inner surface of the tuberosity of the superior maxilla. This surface is rough in all parts except at a central groove running forwards, which combines with a similar depression on the maxillary surface to form the palatine canal. Behind this the bone is smooth; it is over here the superior maxillary division of the sixth nerve and the internal maxillary artery course in their way from the sphenoidal to the maxillary hiatus. The extreme posterior part of the inferior margin articulates with the crus of os sphenoides, assisting to form the pterygoid ridge, to the superior part of the internal surface of which is attached the pterygoid bone.

OS PTERYGOIDEUM is a small plate of bone attached to the under surface of the cranium at its junction with the face; consists of two parts, the anterior of which is flattened from side to side, the posterior from above downwards and is the widest and thinnest. The *anterior* forms a small tuberous prominence, attached to the inner surface of the palatine bone, curved slightly outwards at its extremity, thus producing a groove through which tensor palati plays (bound down by the pterygoid ligament). The superior surface of the posterior part articulates with ossa sphenoides and palati, and its inner margin meets the vomer. To its under surface masseter internus and pterygoideus gain either direct or indirect attachment.

OS ZYGOMATICUM OR THE MALAR BONE forms the major portion of the outer boundary of the orbit. It is pyramidal, presenting three surfaces, three margins, a base and an apex. The *external surface* is for the most part convex, anteriorly it is flat, affording attachment to nasalis longus labii superioris. Its *inferior margin* is surmounted by the middle portion of the zygomatic ridge, to which masseter externus is attached. The *superior surface* is concave, occupies the centre of the bone, being continued neither to the superior extremity nor to the inferior, and assists in forming the orbit. The *internal surface* is concave, and wholly occupied by a squamous surface; the major portion lies in apposition with the superior maxillary bone, but at the antero-superior angle it unites with the lachrymal bone. The *base, external and internal margins* are the sharpened edges of the squamous surface. The *superior margin* anteriorly is serrated for union with os lachrymale, centrally smooth, rounded, and concave, forming the outer margin of the orbit; posteriorly it is continuous at the *apex* on to the superior surface, forming a squamous facet for the antero-external part of the zygomatic process of the squamous temporal bone.

The **VOMER**, so named from the resemblance of the corresponding bone in man to a ploughshare, is the only single bone in the facial portion of the upper jaw. It presents two margins, two extremities, and two surfaces. This bone posteriorly forms the line of division of the posterior nares; anteriorly it is continued almost as far as the anterior naris on the floor of the nasal cavity between the two chambers. Its *inferior border* anteriorly presents a roughened articulatory surface, which fits into a groove formed by the junction of the palatine processes of the superior and anterior maxillæ of one side with the corresponding parts of the other, being an example of the form of joint known as schindylesis. The posterior part of th' inferior border is covered by mucous membrane; centrally it is sharp, dividing the posterior nares, posteriorly expands, terminating on either side in a sharp point, which is concealed by the pterygoid bone. The *superior margin* consists of two thin plates of bone passing upwards one on each side to form a groove into which septum nasi fits. These plates gradually increase in length (and the groove accordingly in depth) from before backwards to about the anterior part of the posterior third of the bone, where they become widened, presenting on their external surfaces concave squamous articulations for the inner ridges of the palatine bones. At the extreme posterior part they meet the under surface of the ethmoid and sometimes extend as far as os sphenoides. The groove also becomes wider here, receiving the postero-inferior thickened angle of septum nasi. The *apex of the bone* rests on the palatine processes of the anterior maxillæ. The *base* is formed by the posterior angular prolongations with their separating depression. It consists of a sharp concave ridge. The *surfaces* are the outer surfaces of the thin plates, and are covered by the Schneiderian mucous membrane.

THE LOWER JAW

is formed by a single bone, the **INFERIOR MAXILLA**, which articulates with the upper jaw only at the glenoid cavities of the squamous temporal bones, though connected with it by soft structures in many parts in the fresh subject. It consists of two exactly similar halves united together by the symphysis at the interior part. Between these anteriorly is situated the tongue, postero-superiorly the pharynx and guttural pouches. Each side of the bone presents two surfaces, two margins, and two extremities. The *superior extremity* is divided into two parts by a deep notch through which a branch of nerve passes.

The anterior part is the **coracoid process**, which runs upwards for a considerable distance, inclined backwards. It is flattened from side to side and rests in the temporal fossa. Its inner surface affords attachment to temporalis. The posterior part is the **condyle**; it is elongated from side to side, convex in every direction, with a slight oblique inclination inwards. It is coated with articular cartilage, and through the medium of the inter-articular disc of cartilage, articulates with the glenoid cavity of the squamous temporal bone. The condyle affords attachment to the capsular ligament around its margin, externally also to its strengthening accessory band. Below the condyle on the inner side pterygoideus is inserted, and here the external carotid artery terminates its temporal branch, winding round the posterior margin of the jaw. The *anterior extremity* or **body** is flattened from above downwards, presenting two surfaces and two margins. The *superior surface* is concave, sloping from without inwards, covered by the mucous membrane of the mouth, from which internally frænum linguæ passes. The *inferior surface* is convex, sloping from the external to the internal margin and backwards.

Superiorly it is covered by the gums; below this depressor labii inferioris has its origin. The *external margin* presents three alveolar cavities for the lower incisors, slightly smaller than those in the upper jaw, and behind these, after a short interval, *the alveolar cavity for the tush*, which is more anteriorly placed than its superior corresponding cavity. The *internal margin* is rough, and by uniting with its fellow forms the **symphysis**. Behind the body the bone is constricted, forming the **neck**, which partly lies in contact with the buccal mucous membrane. Its *external surface* superiorly presents the curved line of attachment of carinus; below this is the *anterior maxillary foramen or foramen menti*. Below this the bone is subcutaneous. Both its *superior* and its *inferior margins* are rounded. The latter to its inner side has digastricus attached as far forward as the symphysis. The neck terminates posteriorly where the *alveolar cavities of the molars* commence. They resemble in general characters and number those of the upper jaw, and occupy a considerable portion of the superior margin of the bone, which extends gradually backwards and upwards, and which from the last alveolar cavity curves upwards to join the anterior margin of the coracoid process, affording attachment to bucciuator and to retractor labii inferioris. Buccinator also becomes attached to the bone up against the alveolar cavities on the external surface; below this retractor labii inferioris runs forwards, the rest of the anterior part of this surface is separated from the skin only by retractor anguli oris. But behind the molars the bone becomes much expanded, in consequence of increased convexity of the inferior margin, which protrudes postero-inferiorly forming the **angle of the jaw** to which sterno-and stylo-maxillaris are attached, the latter to the inner, the former to the outer edge. The margin above this is thickened and tuberous. The parotid gland is in contact with it. The expanded portion is termed the **ramus or branch**, and being concave both externally and internally its central portion is very thin. On both surfaces it presents roughened ridges extending from above downwards which on the external surface serve to give attachment to masseter externus, on the inner side to masseter internus, which latter muscle hides the posterior maxillary foramen to which the inferior dental artery and nerve pass, and having entered it course along below the fangs of the molars as far as the anterior maxillary foramen, where each sends a branch outwards, while the main portion continues on in the bone to supply the incisors and tush. The upper edge of the internal surface up against the molar teeth is termed the **alveolar ridge**, and affords attachment to mylo-hyoideus, the external surface of which muscle lies in contact with the bone as far down as the inferior margin, where the submental branch of the submaxillary artery passes forwards, just above the attachment of digastricus. Below this the inferior margin of the bone is rounded and covered with panniculus; at the commencement of the ramus it presents one or more prominences, which serve to mark where the submaxillary vessels and the parotid duct wind round the jaw.

THE VERTEBRÆ.

The distinctive characteristic of vertebrata is the possession of a central chain of bones forming a **spinal column**. The chain is composed of **vertebræ**, and to the vertebræ all the other bones of the body are appended, so that each vertebra with its bony appendages constitutes a *vertebral segment of the body*. These vertebral segments differ in number in different animals, and frequently vary in animals of the same species. They are between fifty and sixty in number in the horse. **A complete or typical vertebral segment** consists of two similar arches united in forming a rounded **body or centrum**. One is placed superiorly to the centrum, and as it encloses the nerve-centres of animal life is termed the *neural arch;* the other is inferiorly placed, and is termed the *hæmal arch,*—it encloses the organs of vegetative life, including circulation, respiration, digestion, and involuntary motion. Each arch may be divided into five parts; those of the neural arch are: *pedicles*, running upwards from the supero-lateral parts of the body; *laminæ*, running from the upper extremities of the pedicles to meet in the central line, and from the point of junction the *neural or superior spinous process* runs upwards. These parts may be found in almost all the vertebral segments, though only in a rudimentary condition in the coccygeal region. The hæmal arch can be recognised only in the dorsal, though found under extreme modification in most of the other regions. Here we see the *ribs* running downwards from the lateral parts of the body, continued by the *sterno-costal cartilages*, between which, at their lower extremities, is the *sternum or hæmal spine*. From various parts of the arches **processes** project; thus at the anterior and posterior extremity of each lamina is an *articular or oblique process*, with a synovial facet for union with the neighbouring vertebræ. A process is also found on the point of union of the neural laminæ and pedicles which generally unites with another from the point of junction of the pedicles and the centrum, forming the *transverse processes of the cervical and dorsal vertebræ*. The *lumbar transverse processes* consist simply of the inferior process, the superior in this region being blended with the anterior oblique. The two in the cervical region are distinct at their commencement, but united at their free extremities; thus they form the vertebral foramina. From the under surface of the centrum, in some parts, a process arises and passes in a downward direction; this is generally termed the *inferior spinous process*.

The **spinal vertebræ**, as a rule, present the centrum with the neural arch and its processes. The **centrum** has a flattened *superior surface*, from each extremity of which triangular roughened surfaces extend towards the centre, where their apices are slightly separated by a smooth channel which connects other channels situated between the roughened surfaces and the inferior extremities of the laminæ. The *anterior surface* of the body is convex, and is attached by means of a disc of fibro-cartilage to the *posterior surface* of the vertebra in front, which is correspondingly concave. The extremities of the body are its widest parts, so that its central portion is smallest; it presents two surfaces converging to form a prominent line extending from before backwards along the *under surface*. It is this line which, in some cases, forms the *inferior spinous process*. The **ring or arch** of the vertebra is formed by the *laminæ and pedicles*; the latter present grooves on their anterior and posterior margins, and the anterior groove of the vertebra, uniting with the posterior groove of the vertebra in front of it, forms an *intervertebral foramen or gap*, by means of which nerves escape from, and vessels pass into, the spinal canal. The vertebræ of the spinal column may be *true or false*, the **true vertebræ**, or those which are capable of more or less extensive motion upon the neighbouring vertebræ, and which present a regular set of the processes of the neural arch, including spinous (superior and inferior), oblique (anterior and posterior), transverse (simple or compound). True vertebræ are found in the cervical, dorsal, and lumbar regions; they do not all present *all* the processes above named, but their processes present a certain amount of regularity. This is different in the coccygeal vertebræ, and as the sacral bones coalesce the vertebræ in these two regions are termed **false**. The distinction between true and false vertebræ is very artificial. The vertebræ in these different regions present peculiarities which are distinctive, but it will be observed that the characters are gradually assumed; thus, the vertebræ in the posterior part of a region tend to assume the figure of those of the anterior part of the next. The specific characters are, therefore, best marked in the central vertebræ of a region. We shall, therefore, describe the third cervical, ninth dorsal, and third lumbar vertebræ as typical of their several regions. The sacrum will be examined as a whole, the coccygeal vertebræ will require but slight notice.

THIRD CERVICAL VERTEBRA

presents a regular ring and set of processes. The **body** is long, enlarged at each extremity. The *anterior extremity* is heart-shaped, looks downwards and forwards, is convex, and has its base superiorly placed. It is convex in all directions, and by a disc of cartilage is connected to the posterior surface of the body of the second cervical vertebræ. The *posterior extremity* is mainly occupied by a large concave surface, to which the intervertebral disc is attached. Its margin is slightly flat superiorly, and it looks upwards and backwards. Its outer circumferent margin is a sharp ridge; its upward aspect serves to prevent dislocation downwards. The *upper surface* of the body presents the arrangement already noticed, and through the lateral and the transverse grooves run venous sinuses. The roughened surfaces are for attachment of the superior vertebral ligament, and are large in corresponding proportion to the size of the body. The *lateral surfaces* of the body are somewhat rough, and converge towards the central line of the inferior part where they meet, in forming a ridge extending from the apex of the anterior articular surface, to a roughened tubercle situated below the centre of the posterior articulatory surface. This is the **inferior spinous processes,** and to them, as well as to the lateral surfaces, which are slightly concave, longus colli is attached. The **pedicles** present the superior root of the transverse process arising from their external surface, extending obliquely from the anterior oblique process downwards and backwards. The **superior spinous process** arises as a sharp ridge, becoming expanded, rough, and blended with the posterior oblique processes posteriorly; to it are attached ligamentum nuchæ and the superior spinous ligament. On either side of this the *superior surface* of the **laminæ** is concave and rough, anteriorly presenting a groove circumscribing the anterior oblique process, posteriorly a circular line, marking the commencement of the roughened upper surface of the posterior oblique process. The *anterior and posterior margins* of the laminæ between the oblique processes are concave, and afford attachment to the interlaminal ligaments. The **oblique processes** are all very large, present extensive roughened surfaces for muscular attachment, and large circular articulatory surfaces. The anterior articulatory facets look upwards and inwards, and are slightly convex, the posterior downwards and outwards, and are slightly concave. The two anterior oblique processes are connected by the above-mentioned sharp anterior margin of the laminæ, the posterior processes by a thick concave roughened line, which is the posterior margin of the laminæ, and serves to unite two roughened prominences separated from the articulatory surfaces by a slight groove. The **transverse processes** each present two roots separated by a canal, through which the vertebral artery vein and branch of the sympathetic nerve pass. The *upper root* is derived from the outer surface of the pedicle at a line extending obliquely from its antero-superior to its postero-inferior angle. The *other root* commences at the supero-lateral part of the body of the vertebra. The *vertebral canal* is continued backwards to the posterior part of the body by a smooth groove; anteriorly it commences at a funnel-shaped opening just behind the anterior articulatory surface. The process has *two tubercles* on its free margin, both of which are rough and tuberous. The anterior extends obliquely downwards and forwards, the posterior runs horizontally backwards, slightly curled upwards at its roughened free margin. Both of these tubercles afford attachment to muscles. The *inferior surface* of the process is concave and rough; the *superior surface* convex. Its *anterior margin* is concave, smooth, and rounded; its *posterior margin* concave and sharp.

CERVICAL VERTEBRÆ

in general are remarkable for their size, for the large size of their foramina magna and intervertebral gaps, for the regularity and uniform length of their processes, and for their general roughnesses, and the facilities they afford for muscular attachment. Each of them, however, presents modifications. The first two, atlas and dentata, approximate the characters of the vertebræ which are modified to form the skull; the last two tend to assume the dorsal type. The third, fourth, and fifth are with difficulty distinguishable. *The following modifications will be observed in examining the vertebræ from the third to the seventh.*

The posterior part of the superior spinous processes becomes more separated from the posterior oblique processes. The anterior and posterior oblique processes in the third are separate, in fourth united by a thin ridge, in fifth by a roughened ridge. The concavity between the superior spinous process and the anterior oblique processes becomes more and more marked to the sixth. The articulatory surfaces on both oblique processes increase in size from before backwards, as do also the vertebral foramina, and the articulatory surfaces gradually change from the vertical to the horizontal position. The inferior spinous process decreases in size to the sixth, where it is almost imperceptible; it becomes larger in the seventh and very irregular. The posterior tubercle from before backwards manifests a tendency to division, which in the sixth is very marked, so that the superior and inferior parts form distinct prominences, of which the superior is triangular, concave posteriorly. From the dentata to the sixth the superior spinous processes decrease in size, but the seventh has a broad superior spinous process terminating superiorly in a point. This is the shortest of the cervical vertebræ. Its transverse process is square, undivided, and has only one root. Its posterior oblique processes present much smaller articulatory surfaces than the anterior, but not the smallest in the neck. They come into contact with the anterior oblique processes of the first dorsal vertebra. This vertebra, at the supero-posterior and lateral parts of the body, has a concave synovial facet on each side for the head of the first ribs. The first and second cervical vertebræ require more detailed description.

ATLAS is the first cervical vertebra, and in it we see a manifestation of analogy with the cephalic vertebræ in the expansion of its parts. Its **ring** is very large, ovoid from side to side anteriorly, circular posteriorly. It accommodates the commencement portion of the spinal cord with its membranes, the commencement portion of the basilar and of some meningeal arteries, the spinal accessory nerve, some venous sinuses and fat. Also posteriorly *the odontoid process of the dentata*, with its ligaments. The **body** is very small, flattened from above downwards. Its *superior surface* anteriorly has a roughened groove, extending on either side outwards and backwards for attachment of the odontoid ligaments. Posteriorly it presents an articulatory surface concave from side to side, surrounded by a ridge, continuous with the articulatory surface upon the posterior part of the bone. The *inferior surface* is convex. Anteriorly it has the lower part of the articulatory surfaces for the condyles of the occiput, converging inferiorly, where they are separated by a groove to which their synovial membranes are attached, and which is continuous with a small triangular space on the antero-inferior part, which is occupied by a ligament. At the postero-inferior part of the body is a prominence looking backwards, **tubercle or inferior spinous process**, to which longus colli (and the inferior vertebral ligament) is attached. On either side of this the bone is rough for attachment of obliquus capitis anticus. Opposite the articulatory surfaces it is convex and bulging, between these somewhat concave and smooth. The **pedicles** run obliquely upwards and outwards from the lateral parts of the body of the bone, being concave and rough *internally*, convex *externally*, and expanded at either extremity, forming those portions of this vertebra which correspond to oblique processes of others. Anteriorly the expanded part presents an articulatory facet for union with a condyle of the occiput. It superiorly looks downwards and slightly forwards, running outwards. It then becomes reflected inwards and looks upwards; inferiorly, a small portion looks directly inwards and converges towards, but does not quite meet, its fellow. Thus we have formed an irregular cotyloid cavity.

The posterior expanded portion of the pedicle presents a convex triangular facet, having its apex at the postero-external angle of the lamina sharp and acute, its infero-external angle rounded, its infero-internal angle continuous with the smooth postero-superior part of the body. It looks inwards and backwards, and comes into connection with the anterior surface of the body and with the odontoid process of the dentata. The **laminæ** *internally* are concave and rough; *externally* convex, having a small ridge centrally at the anterior part, a rounded and roughened prominence for attachment of the superior spinous ligament and rectus capitis posticus minor posteriorly. The anterior margin is a sharp ridge, forming a concavity affording attachment to the capsular and lateral ligaments of the attoido-occipital joint, extending from the articular surface for one condyle to that for the other. At the antero-external angle the point of junction of the lamina and pedicles is pierced by a foramen extending into the spinal canal, continuous with one in the anterior part of the wing. This gives passage to the suboccipital nerve, and to a branch of artery going inwards to assist in forming the basilar. The **transverse process ala, or wing**, extends from above downwards and backwards, obliquely to the infero-external angle of the posterior articular surface. It is a thin plate of bone, extending downwards and outwards; since its surfaces are concave it is thinnest at its centre, and here it is pierced by a foramen leading from a groove, through which the vertebral artery passes from above downwards. Its outer margin is convex, thick, and rough, and extends obliquely downwards and outwards, posteriorly a little behind the rest of the bone at its most convex part. It affords attachment to several muscles of the neck, levator humeri, splenius, trachelo-mastoideus, also to obliquus capitis superior and obliquus capitis inferior. In the concavity beneath the wing the union of the vertebral artery with ramus anastomoticus takes place, and the resulting branch passes through the anterior foramen, thus gaining the superior surface, sending one branch into the spinal canal, the other to the poll. The superior cervical ganglion is also situated in the concavity of the under surface of the wing.

The **DENTATA OR AXIS** is the largest and the longest of the true vertebræ, and has the largest superior spinous process; it is the second cervical vertebra. Its **body** is very long, but *superiorly* presents the usual arrangement for venous sinuses and the attachment of the superior vertebral ligament. *Anteriorly* it presents a synovial surface for union with atlas, and centrally the **odontoid process**: superiorly this process is separated from the rest of the upper surface of the body by a rough line from which another runs forward to the anterior margin. The space on either side between the ridges is concave; to the transverse ridge is attached the broad, to the concavities the short odontoid ligaments. The anterior margin is convex, the lateral margins sharp and elevated. The inferolateral surface of the process is convex and articulatory, coming into connection with the postero-superior part of the body of the atlas. It is most prominent at the centre of the posterior part, forming a small tubercle. It is continued on either side on to the antero-external part of the body as a rounded synovial surface, looking obliquely upwards and outwards. The *under surface* of the body centrally presents a prominent ridge, sharp posteriorly, roughened anteriorly. It affords attachment to longus colli and to the inferior vertebral ligament. Anteriorly it does not extend as far as the odontoid process, from which it is separated by a groove, which extends completely round the anterior part, affording attachment to the capsular ligament of the atlo-axoid articulation, separating the inferior spinous ridge from the outer circumferent margin of the anterior articulatory surfaces, which here cause expansion of the bone. *Posteriorly* the body laterally presents two surfaces, converging towards the inferior ridge roughened for attachment of longus colli. From the supero-lateral part, of the body the **transverse processes** run obliquely upwards outwards, and backwards; they have two roots, a broad concave one, inferiorly separated by the vertebral foramen from a smaller one superiorly situated. This process is undivided at its free extremity. The **pedicles** are long and high, thus rendering the **foramen magnum** longest vertically; their *inner surface* is rough and concave, their *outer rough* and straight. The *posterior margin* is concave, assisting the third cervical vertebra to form an **intervertebral gap**. The *anterior margin* is very deeply indented, generally forming a large foramen, bounded anteriorly by a small thin process of bone or of white fibrous tissue; from this a smooth broad grove runs downwards to the vertebral foramen; it gives exit to the second cervical nerve, and also gives passage to vessels. The *posterior surface* of the body and the **oblique processes** (of which there are but two, posteriorly placed) resemble those of the following cervical vertebræ already described. The **laminæ** *inferiorly* form the superior part of the ring, *superiorly* are wholly occupied by the **superior spinous process**, which is very thick, and from before backwards increases in height, thickness, and roughness of its superior margin. This margin *anteriorly* affords attachment to rectus capitis posticus major and to the superior spinous ligament, *posteriorly* to ligamentum nuchæ and complexus minor. It bifurcates posteriorly, and its divisions blend with the oblique processes. To the sides of this process obliquus capitis inferior is attached along its whole length.

THE DORSAL VERTEBRÆ

are eighteen in number, and are characterised by the length of their superior spinous processes, and by the synovial articulatory surfaces on the supero-lateral parts of their bodies for union with the heads of the ribs, also by articulatory surfaces on their transverse processes for union with the tubercles of the ribs. They are the smallest of the true vertebræ. In the anterior part they tend to assume the cervical, in the posterior the lumbar type.

NINTH DORSAL VERTEBRA.

The **body**, on the upper surface, presents the usual arrangement. Its two *sides* present superiorly at the anterior and posterior parts concave synovial facets, extending on to the external surfaces of the pedicles. The anterior facet looks forwards and outwards, and comes into connection with the posterior facet of the rib in front. The posterior facet looks backwards, and slightly outwards, and is for the anterior facet of the head of the rib behind. Below this the sides are concave from before backwards, and meet inferiorly in forming a median ridge (which in some of the vertebræ of the dorsal region is elongated to form the *inferior spinous process*). Each of the sides at its centre, though inclined to the anterior part, presents a *medullary foramen*, through which a branch of the intercostal artery passes. The under surface of the bodies of the anterior six dorsal vertebræ is covered by longus colli, of the posterior three by psoæ parvi, of the central ones by the vena azygos on the right side, thoracic duct centrally, posterior aorta on the left. The *anterior surface* is convex and circular, somewhat flattened superiorly. It gives attachment to the intervertebral disc, and is adapted to the posterior surface of the body of the vertebra immediately in front of it. The *posterior surface* is correspondingly concave. The **pedicles** are internally smooth and concave, externally rough, slightly encroached upon by the facets for the heads of the ribs. Anteriorly and posteriorly they have grooves for the intervertebral gaps scarcely perceptible anteriorly, posteriorly, in some cases, converted into a foramen. Centrally, at the point of junction with the lamina, is a tuberous process, rough for muscular attachment (of longissimus dorsi and levatores costarum). This is the **transverse process**, and has on its external surface a small irregular convex synovial facet for the tubercle of the rib, having between it and the posterior facet on the body a smooth groove, extending from the intervertebral gap. The **laminæ** are concave *inferiorly*, forming the roof of **foramen magnum**, *supero-posteriorly*, wholly occupied by the **superior spinous process**, *supero-anteriorly* presenting a small synovial facet on each side, representing the **anterior oblique processes**; the **posterior oblique processes** are on the postero-inferior part. The **superior spinous process**, remarkable for its length, runs obliquely upwards and backwards. *Superiorly* it is tuberous, affording attachment to the superior spinous ligament, and so indirectly to muscles. Its *anterior margin* is sharp, especially inferiorly, affording attachment to the interspinous ligament. It is much longer than the *posterior margin*, which is centrally sharp, having on either side a groove, deepest inferiorly, bounded externally by a roughened margin. They terminate just above the posterior oblique processes. The central sharp ridge bifurcates inferiorly, one division running to each oblique process.

To this margin also is attached the interspinous ligament.

Differences in the Dorsal Vertebræ.

Superior spinous processes increase in length to fifth, then decrease to thirteenth, the rest being about uniform; they are inclined backwards to the sixteenth, which is upright; the seventeenth and eighteenth incline slightly forwards. First comes to a point superiorly; the rest increase in roughness and thickness to the sixth, and in width from before backwards in the twelve posterior vertebræ, in which also may be noticed a concavity just below the anterior part of the superior extremity. In the central vertebræ the anterior margin is rough about at its middle. The *transverse processes* are large and mainly articulatory in the first, and gradually become modified in the last until each of them consists of two portions, the superior of which is blended with the anterior oblique process, the inferior, articulatory, continuous with the anterior facet on the lateral part of the body of the vertebra in the case of the last two. The *articulatory facets for the heads of the ribs* gradually decrease in depth from before backwards. The *intervertebral notches on the posterior part of the pedicles* are in the fourth to the sixteenth vertebræ converted into foramina; hence the fifth to the seventeenth have no anterior notches. The *articulatory surfaces on the anterior oblique processes* are larger than those on the posterior. The former look outwards, the latter inwards; their tendency in this is increased from before backwards. The *oblique processes* of the first four vertebræ are distinct; in the other vertebræ the articular surfaces are continued on to the laminæ. *Inferior spinous processes* are marked only in the six anterior and three posterior vertebræ. The *foramina magna* are largest in the anterior, smallest in the central vertebræ. The *smooth grooves on the upper surface of the body* are most marked anteriorly. The **last dorsal vertebra** has no posterior articulatory surfaces for the heads of the ribs, and its anterior surfaces are blended with those of its transverse processes.

THE LUMBAR VERTEBRÆ

are the vertebræ forming the basement structure of the loins, situated between the dorsal and the sacral regions.

In general characters they resemble the dorsal vertebræ; we shall, therefore, only notice their peculiarities. These vertebræ are five or six in number, are larger than the dorsal, and each presents a large foramen magnum and transverse processes; also has no articulatory surfaces for the heads of the ribs. The **superior spinous processes** resemble those of the posterior dorsal; all look forwards; are short and stout, elongated from above downwards, tuberous at their *upper extremities, anteriorly* presenting a sharp margin, *posteriorly* a double margin, both for attachment of the interspinous ligaments. The **oblique processes** tend to assume a peculiar rounded hinge-like character; those on the *posterior part* present a rounded convexity, elongated from behind forwards, looking outwards and downwards. The *anterior oblique processes* have articulatory surfaces looking upwards and inwards. The *intervertebral gaps* are large, the **transverse processes**, formed by a single root, extend directly outwards, those of the anterior vertebræ having a slight inclination backwards, those of the posterior forwards. Their length is characteristic of the vertebræ of this region. They are flattened from above downwards, and of the first five vertebræ are tuberous at their *free extremities*, affording attachment to quadratus lumborum, psoas magnus, and indirectly to some of the abdominal muscles.

The *anterior and posterior margins* are sharp, affording attachment to intertransversales lumborum; the *upper and under surfaces* are smooth; the former affords attachment to longissimus dorsi, which occupies the space between this and the superior spinous process; the latter to the psoæ muscles. When six lumbar vertebræ exist the **transverse process of the sixth** is much shorter than the rest, and presents *articular synovial facets, both on its anterior and its posterior margins*, ovoid and elongated from side to side, separated by grooves from the amphiarthrodial facets of the body. These present special ligaments, and the posterior one slants from without inwards and backwards, being in connection with the first sacral transverse process. The anterior slopes in a similar direction, articulating with a corresponding facet on the posterior part of the fifth lumbar transverse process. The latter, where there are only five lumbar vertebræ, is modified to join the sacrum. Most horses present six, most asses five, lumbar vertebræ. The **bodies** of these vertebræ are larger and more prismatic than those of the dorsal, terminating *inferiorly* in somewhat marked ridges, **inferior spinous processes,** for attachment of the inferior vertebral ligament, to which the crura of the diaphragm are attached. To their *lateral surfaces*, the psoæ parvi muscles are attached; their *anterior and posterior surfaces* present but shallow concavities and convexities. The bodies and foramina magna decrease in height from before backwards; so at the posterior part of the last lumbar the body presents a shallow surface, very much elongated transversely.

THE SACRUM,

in early life, consists of five distinct vertebræ, which in the adult become firmly united by healthy bony union, whereby a firm basis for attachment of the hind limb is afforded. We therefore shall view it as a single bone, since it acts as such. It is situated at the posterior part of the spinal column, between the lumbar and coccygeal vertebræ. It articulates laterally with the venter ilii. It is pyramidal in figure, its *base* being anteriorly placed; sometimes it is composed of six bones, in consequence of coalescence of the first coccygeal vertebra with its *apex*. It is a false vertebra, as its segments are not capable of moving upon each other. The **base** presents laterally **foramen magnum** coming to a point superiorly at the root of the superior spinous process, broad inferiorly. It leads into a canal which gradually and regularly decreases in size from before backwards, being extremely small posteriorly. The spinal cord extends into this about as far as opposite the second sacral bone; the membranes of the cord are continued farther. Below this opening is an *articulatory surface*, the transverse diameter of which is about twice as long as the vertical; it is slightly convex in all directions. At the supero-lateral parts of the foramen are the **oblique processes**, which are roughened externally, and internally present concave facets looking upwards and inwards. From the inferior part of the anterior margin of each of these a thin ridge runs to meet its fellow at the root of the superior spinous process of the first bone. Below the oblique processes are the **intervertebral notches,** which lead into grooves running from the upper to the under surface of the transverse process of the first bone, *separating the articulatory surface of the body from those on the anterior margins of the transverse processes*, which overhang the grooves inferiorly in a remarkable manner. These surfaces come into contact with those on the posterior margin of the transverse processes of the last lumbar, of which they are the exact counterparts, being almost flat, looking inwards and slightly upwards. They extend to the outer extremity of this transverse process, which is sharp, and at its superior surface form angles with rounded roughened surfaces which extend to the extremity on the postero-external part, much expanded, looking upwards, backwards, and outwards, for union with the venter surface of the ilium. They do not occupy the whole of the *upper surface*, but are continued by a roughened ridge, which extends along the lateral part of the bone, becoming smaller and sharper posteriorly, formed by union of the transverse processes of the other sacral bones; it affords attachment to the sacro-sciatic ligament. Both above and below this ridge is a series of **foramina**, four in number, decreasing in size from before backwards for passage of the branches of the sacral nerves, the superior branches through the superior foramina, the inferior branches through the inferior foramina, through which also pass branches of the lateral sacral arteries and veins. Above the superior foramina (supersacral) are a series of small prominences, decreasing in size from before backwards, produced by the **oblique processes of the bones.** Higher up are the **superior spinous processes,** more or less completely united by ossification of the interspinous and interlaminal ligaments, presenting a series of five tuberous prominences, to which triceps abductor femoris, biceps rotator tibialis, ischio-tibialis, and some of the coccygeal muscles, also the lateral sacral ligament and longissimus dorsi are attached. The first spinous process looks almost directly upwards; the rest have a marked direction backwards, and decrease in length posteriorly. The *inferior surface* of the bone anteriorly, on either side, presents a remarkable smooth, irregular concavity, being the under surface of the transverse process; a rounded neck, extending inwards and backwards, serves to unite this to the main portion, which is formed by the under surfaces of the bodies, as denoted by transverse, irregular ridges, which serve to divide this portion into five flattened parts.

It is slightly concave from before backwards, is widest anteriorly, and decreases in size posteriorly; on either side it slopes upwards and outwards, presenting the *inferior row of foramina* (subsacral), which are much larger than the superior. Internally placed to the foramina the lateral sacral arteries run along this portion of the bone. The **apex** consists of a somewhat rounded and slightly concave articular surface, surmounted by a small ring, presenting on either side a small transverse, superiorly a superior spinous process. It articulates with the first coccygeal vertebra.

THE COCCYGEAL VERTEBRÆ,

from thirteen to eighteen in number, occupy the posterior part of the vertebral column, and are small bones in which the different parts of a vertebra assume a more and more rudimentary form from before backwards, so that while the *first* presents a body, ring, and small processes, the *sixth* presents only a body with two pedicles and small transverse processes; the *posterior coccygeal bones* are merely small bony masses, elongated from before backwards, enlarged at each extremity, where they become united with their fellows by fibro-cartilage. The posterior bones only become ossified late in life, so that the number of these vertebræ varies with age. They all present both anteriorly and posteriorly *convex* articulatory surfaces to their bodies. In consequence of this, mobility of the vertebral column here attains its maximum, but the want of a regular set of processes renders these false vertebræ; superiorly they afford attachment to the levatores coccygis; inferiorly to the depressores; laterally to the curvatores. The anterior bones also afford attachment to ischio-tibialis and to compressor coccygis.

THE RIBS, COSTÆ,

together with the sternum and the cartilages of elongation form the hæmal arch of the dorsal vertebral segments. They are elongated bones, extending from above downwards from the lateral parts of the dorsal vertebræ. Though they have a longitudinal axis greatly exceeding either of the transverse axes, yet as they present no medullary canal nor marked distinction of shaft and extremities, they are not included in the class "long round bones." Each rib presents a body and an upper extremity. The **extremity** is divided into *head*, connected to the rest of the bone by the *neck*, and *tubercle*. The **head** is a more or less spherical process, with two convex, synovial, articulatory surfaces, separated by a roughened groove. The *anterior surface* articulates with the posterior facet on the body of the vertebra in front; the *posterior*, with the anterior facet of the vertebra behind; the *groove* corresponds to the intervertebral disc of cartilage and affords attachment to ligamentum teres. Around each of the articulatory surfaces is a slight groove for attachment of a capsular ligament. In the two last ribs the posterior articulatory surface is continuous with the facet of the tubercle. The **neck** varies in thickness and rotundity in the different ribs; its inner surface is roughened for attachment of the stellate ligament. The **tubercle** seems to be the superior part of the body, and is posteriorly placed to the head. That of the first rib is largest; it gradually decreases in size in the bones from before backwards, so that it is extremely small in the two last. From this process arises levator costæ, and to it is attached longissimus dorsi. It presents on its internal surface, becoming more anteriorly placed in the posterior vertebræ, a small articulatory facet for union with the transverse process of the dorsal vertebræ; the outer circumferent margin of this is rough for attachment of the capsular ligament, and on either side for the costo-transverse ligaments. The **body** is more or less flattened irregularly from without inwards, and in the central ribs is so twisted upon itself that the two extremities will not rest upon a plane surface at the same time. The eighth rib takes the following course from its upper extremity: at first, outwards, backwards, and downwards; centrally, more directly downwards; inferiorly slightly inwards and backwards. The body therefore presents two margins and two surfaces. The *internal surface* is smooth, convex from before backwards, concave from above downwards. It is covered by the pleura costalis, through which it may be seen on laying open the thorax of a fresh subject. At its extreme superior part the sympathetic gangliated cord may be seen. To the extreme inferior part of this surface of the posterior ribs are attached the diaphragm and transversalis abdominis. The *external surface* is convex in all directions; in the anterior ribs its lower part affords attachment to serratus magnus, in the fourteen posterior ribs to obliquus abdominis externus, in the four or five posterior ribs to obliquus abdominis internus. To about the inferior part of the superior third of the last two ribs a line of prominences extend, one on each rib, from the tubercle of the second; to this transversalis costarum is attached, while a corresponding superior line of prominences runs from the fourth tubercle to meet the former in the sixteenth rib; it marks out the inferior limit of longissimus dorsi, and opposite to it the bone makes its most marked curve downwards, which is termed the **angle of the rib**. Superficialis costarum is attached below the inferior row of prominences. The *anterior margin* affords attachment to the intercostales externi at its outer edge, superiorly to levatores costarum, at its inner edge to intercostales interni. In the fourteenth rib a small groove runs along this margin from the head of the bone; it gradually increases in length in the ribs in front of this, so that in the second, third, and fourth the external margin seems bevelled off from behind forwards, affording extensive surface for attachment. The *posterior margin* presents a corresponding arrangement on the inner side. It affords attachment to the intercostales in a similar manner to the anterior margin. Towards their inferior extremities the ribs tend to become rounded, the ninth being most so. The posterior ribs assume the character of thin long bones composed of two rounded surfaces meeting by sharp margins. They increase in width from before backwards to the sixth, after which they decrease. In length they decrease from the eighth both backwards and forwards, the first being the shortest. The *inferior extremity of the body* is continuous with the **cartilage of elongation**, the union being more or less gomphotic. The cartilage has a considerable tendency to undergo ossification, thus assuming the normal character of the inferior costa as found in birds; the bone therefore extends downwards in the centre of the cartilage to a degree varying with the age of the animals. Some of these cartilages are directly attached to the sternum, others indirectly through the medium of their fellows. Those of the first class belong to the first eight ribs, which therefore are said to be **true ribs**, behind these, on each side, are generally ten **false ribs**, sometimes nine or eleven, with the inferior extremity of the cartilage pointed, and more or less firmly united to the neighbouring cartilages by a small elastic ligament. The union of the posterior cartilages is most lax, and in some cases the last rib is wholly unconnected to the last but one; it is then said to be *floating*. The floating rib is generally found when nineteen ribs exist on each side; sometimes it is a mere cartilaginous elongated body extending from the transverse process of the first lumbar vertebra imbedded in the abdominal muscles. The external surface of the fourteen posterior cartilages is connected with the external oblique abdominal muscle. To that of the four or five posterior true ribs rectus abdominis is attached, and anteriorly is blended with lateralis sterni, which is continued obliquely forwards and upwards over the second and third to become attached to the first, from which also sternothyro-hyoideus partly derives its inferior attachment. The cartilages decrease in thickness from before backwards; they increase in length from behind and from before to the ninth and tenth. They are connected together by the inferior thickened portion of the intercostales interni, for the external intercostales terminate opposite the inferior extremity of the rib. The *inferior extremity* of the cartilages of the false and floating ribs terminates in a point, that of the true ribs is convex, coated with articular cartilage, articulating with the cartilage of the sternum.

The **first rib** is the shortest and strongest, and is also the straightest. Its *head* articulates with the last cervical and first dorsal vertebræ. Its *tubercle* is large and very rough externally for attachment of transversalis costarum, and longissimus dorsi. It is irregularly rounded, to the *inner surface* is attached the pleura at its reflection to form the mediastinum. And many of the important structures which pass through the entrance of the thorax lie in contact with this bone. The *posterior margin* affords attachment to the first intercostals. The *anterior margin* presents at the inferior part of the superior third, a prominence to which the short or upper head of scalenius is attached. Below this the bone is smooth for the passage of the axillary plexus of nerves. On the *external surface* at the superior part of the inferior third the lower head of scalenius is attached, and under it are two smooth grooves for the passage of the axillary artery, and (below it) the axillary vein. The rest of this surface is occupied by the attachment of serratus magnus. The cartilage of the first rib is the shortest and stoutest, and very closely approximates its fellow of the opposite side.

The **last rib** is the smallest, and its *posterior margin* gives attachment to quadratus lumborum. To the *inner surface* of the last two ribs the psoas magnus attached. We have seen that the anterior ribs are strongest, broadest, and most firmly connected together; they are thus admirably adapted to the formation of the **thorax**, which not only serves to protect important viscera, but also affords an extensive base for attachment of the fore limbs to the trunk. This cavity is completed below by the **STERNUM**, which consists of the hæmal spines of six vertebral segments embedded in a cartilaginous mass, so that they may be viewed as one bone. Each of these bones presents seven surfaces. The *superior surface* is smooth, forms the floor of the thorax, where it gives attachment to ligaments, and to the sterno-costales interni. That of the sixth bone on either side presents the articulatory surfaces for the cartilages of the last two true ribs, of the fifth bone for the sixth sterno-costal cartilage. The *anterior* and *posterior surfaces* are rough for attachment of the cartilage, while the *superior* and *inferior lateral* surfaces, scarcely distinguishable in the first bone, assist in forming the lateral and inferior surfaces of the bone taken as a whole. These bones decrease in depth and increase in width from before backwards. They are prolonged anteriorly and posteriorly by cartilages. The **anterior cartilage** presents a rounded supero-anterior extremity from which a short concave superior margin runs to the antero-superior part of the first bone. An inferior margin convex, rounded smooth along the central line of the under surface, blending with the inferior margins of the bone almost as far backwards as the posterior cartilage. It is flattened from side to side, though growing broader posteriorly, and thus it gives the sternum a keel-shape; it is the **cariniform cartilage**. Its prominent antero-superior part would form the prow. The **ensiform, posterior or xiphoid cartilage** is a membraniform piece of cartilage, flattened from above downwards, widest at its attachment to the posterior part of the last bone, decreasing in size posteriorly to form the *neck*, subsequently again expanded in forming an *appendix portion*. To its under surface are attached the abdominal oblique and rectus muscles, also pectoralis magnus, to the upper surface the diaphragm and transversalis abdominis and the fibrous pericardium. The sternum thus presents three surfaces and two extremities. The *anterior extremity* is the prow of the cariniform cartilage; to it sterno-maxillaris and sterno-thyro-hyoideus are attached. The *posterior extremity* is formed by the ensiform cartilage. The *lateral surfaces* anteriorly give attachment to pectoralis anticus, posteriorly to pectoralis magnus and transversus. The *inferior margin* is anteriorly smooth and rounded; it here presents a spurious bursa, over which pectoralis transversus plays. The *upper surface* is slightly concave, both longitudinally and transversely. Centrally it affords attachment to the mediastinum. Anteriorly the thymus gland rests on it, posteriorly the fibrous pericardium is attached. Laterally it presents the articulatory surfaces for the inferior extremities of the sterno-costal cartilages, and internally placed to this the internal thoracic artery runs from before backwards, with its accompanying vein. The first-sterno-costal cartilage articulates with the sternum just in front of the first bone, the second, third, fourth, and fifth with the intervals between the following bones, the sixth with the fifth bone, the seventh and eighth with the last.

BONES OF THE FORE LIMB.

The **SCAPULA** is a flat triangular bone, placed on the antero-lateral part of the thorax, having its apex inferiorly opposite the first rib, its base extending obliquely downwards as far as the angle of the sixth rib. A *flat* bone is required in this position to afford sufficient surface for attachment of the large muscles which connect the fore limb to the trunk, for in the horse two bones (coracoid and clavicle) which perform this function in many other animals are rudimentary. It presents two surfaces, three borders, and three angles. The *anterior border* is superiorly convex and roughened, inferiorly concave and smooth, and at its extreme inferior part presents a large roughened process, **coracoid** (a rudiment of a distinct bone). This process *externally* presents, superiorly a groove for a blood-vessel running obliquely downwards and backwards; below, and almost parallel to this a roughened ridge, mapping out the line of attachment of flexor brachii, may be traced in a circular manner around the whole anterior part of the process. On the *inner surface* is a ridge with a slight inclination backwards, terminating inferiorly in a *small process* to which coraco-humeralis is attached. Around the anterior margin of the bone, just above the coracoid process, runs arteria dorsalis scapulæ with its accompanying vein and nerve from within outwards, protected by *a band of white fibrous tissue* (*coraco-scapular ligament*), which extends from the process to about the middle of this margin of the bone; antea spinatus becomes attached to this ligament externally; subscapularis internally. The roughened anterior margin at its superior part affords attachment to pectoralis anticus and antea spinatus. The *postero-superior* margin of the bone is rough and cancellous for attachment of the **scapular cartilage of elongation.** This is a thin piece of fibro-cartilage, extending upwards for about three inches, having a tendency to ossification at its *attached margin*. It becomes very thin *superiorly*, and curls inwards. *Anteriorly* its margin has a backward inclination, *posteriorly* it is continued as a thin piece of cartilage beyond the posterior angle of the bone. Its *external surface* affords attachment, anteriorly to antea spinatus, centrally to postea spinatus, and posteriorly to serratus magnus. To its *inner surface* the rhomboidei are attached. The postero-superior margin anteriorly forms an obtuse angle with the anterior margin. Posteriorly, in a corresponding manner, it joins the postero-inferior margin, at the extreme superior part of which is a triangular surface roughened for attachment of serratus magnus. From this the *posterior margin* of the bone makes a curve in a downward direction to the inferior angle. Superiorly this curve presents a sharp edge, to which teres externus and internus, caput magnum of the triceps extensor brachii and scapulo-ulnaris are attached. Inferiorly it is more rounded, but still, for the most part, roughened; centrally for attachment of caput magnum, around which the attachment of scapulo-humeralis extends in a peculiar manner, at the extreme inferior part of scapulo-humeralis posticus, externally of scapulo-humeralis externus, internally of subscapularis. About four inches from its inferior part, in some cases, we find a groove running on to the external surface of the bone, marking the course of branches of the posterior scapular vessels. The portion of the bone opposite this is more rounded than the rest, and is termed the **neck**. Below it are the coracoid process anteriorly, the articulatory surface for the humerus posteriorly. This *articulatory surface* occupies the **inferior angle**. It is termed the **glenoid cavity**, and is much smaller than the head of the humerus on which it fits, thus allowing great range of action of the joint. It looks slightly downwards and backwards, is shallow, and presents a sharp margin, but at its antero-external part manifests a slight tendency to extend on to the external surface of the bone. At its anterior part it is divided into two by a groove into which, when the limb is extremely bent, coraco-humeralis tendon may fit. Externally to this is a roughened prominence, near the articulatory surface. The rest of the space between the coracoid process and the glenoid cavity is smooth, and presents a large foramen for a blood-vessel to pass into the bone. The sharp outer circumferent margin of the glenoid cavity affords attachment to the capsular ligament of the shoulder-joint.

The *external surface* of the scapula is termed the **dorsum**. It is convex in all directions, but is hollowed out in some parts in such a manner as to produce a very prominent ridge, extending from the superior margin to the neck, separating the anterior third from the posterior two thirds of this surface. This is the **spine**, and separates this surface into *fossa antea spinatus* anteriorly and *fossa postea spinatus* posteriorly. At its attached margin, and superiorly and inferiorly, this spine gradually blends with the surface of the bone. Its free margin presents inferiorly a roughened prominence (corresponding to the *acromion process* in man with which the clavicle articulates). More superiorly the ridge presents a roughness which culminates in an elongated prominence with a backward curve, the **tubercle**, to which the inferior angle of trapezius is attached. The spine anteriorly affords attachment to antea spinatus, posteriorly to postea spinatus ; also levator humeri, teres externus, and (indirectly) pectoralis anticus gain attachment to its edge. **Fossa antea spinatus** is bounded anteriorly by the anterior margin, posteriorly, for the most part, by the spine, superiorly by the postero-superior margin, inferiorly it becomes widened in forming the anterior part of the neck; to its whole extent is attached antea spinatus. This surface is increased by the addition of the space, bounded anteriorly by the *coracoscapular ligament*, under which the dorsalis scapulæ vessels and nerve wind round the neck of the bone in passing to or from this fossa. **Fossa postea spinatus** occupies two thirds of the dorsum. It is bounded posteriorly by the postero-inferior margin, superiorly by the postero-superior margin, anteriorly by the spine. Superiorly it is smooth, affording attachment to postea spinatus muscle, while the inferior part, forming the posterior part of the neck, presents roughened lines extending obliquely upwards and backwards from the inferior extremity of the spine. This part gives attachment to scapulo-humeralis externus, and presents grooves running from the posterior margin of the bone to the inferior extremity of the spine, in which a branch of the posterior scapular artery runs, sending off *the medullary artery of the bone through a foramen in this region*. The *inner surface* of the scapula is termed the **venter**. It is concave from before backwards, the concavity being most marked along the central line opposite the spine. It is widest centrally, and superiorly terminates in a point on reaching the cartilage of elongation. Inferiorly it becomes flat in forming the neck, and again concave at the inner surface of the coracoid process. Across the neck of the bone, from behind forwards, run branching grooves for arteries; in other respects this concavity is smooth, and it is continuous at its upper part, anteriorly and posteriorly with two other smooth surfaces, the anterior of which is smallest and bounded by a roughened line running downwards and backwards from the anterior edge of the bone; the posterior is bounded by a curved and much rougher line extending from the posterior margin of the bone up against the posterior angle. This smooth surface, therefore, presents superiorly three points; inferiorly it terminates at the margin of the glenoid cavity, the inner surface of the coracoid process, and posteriorly the roughened line which affords attachment to scapulo-humeralis externus. It affords attachment to subscapularis, between which, therefore, superiorly are two roughened triangular surfaces, to which are attached the anterior and posterior parts of serratus magnus. Along the inner edge of the postero-superior roughened margin, and to the inner surface of the cartilage of elongation, are attached the rhomboidei muscles.

HUMERUS is a long round bone, which is situated below the scapula, running obliquely from above downwards and backwards, forming with the scapula an obtuse angle posteriorly. It is a very stout bone, and its superior extremity is much larger than its inferior. The *superior extremity* presents a head and four tubercles. The latter form a continuous chain occupying the anterior part of this extremity. The head is situated at the posterior part and is convex. Since it looks directly upwards in the natural position of the bone it seems to be inclined to the posterior part. It extends slightly on to the fourth tubercle, and between it and the other tubercles is a space elongated from side to side with numerous large foramina; this is occupied by fat in the fresh subject. On the inner side of the upper extremity is a roughened process, broad posteriorly, terminating in a point anteriorly in joining the internal tubercle; to this subscapularis is attached. The three tubercles on the anterior part are termed respectively external, middle, and internal. The *internal tubercle* is the smallest, receiving the termination of the roughened spot just mentioned, posteriorly, and externally presents the commencement of a synovial surface coated with fibro-cartilage which dips down into the shallow fossa between the internal and middle tubercles and then, after completely covering the *middle tubercle*, dips into the deep fossa between this and the *external tubercle*, terminating after investing the inner surface of the latter. Over this surface the tendon of flexor brachii plays; hence it is lubricated with synovia, and presents a large and important bursa, which is separated from the capsular ligament of the joint by the fat in the perforated space in front of the head. Below this surface may be seen a series of small articular foramina. Both the external and internal tubercles afford attachment to antea spinatus and pectoralis magnus which serve to bind flexor brachii in its situation, but the external tubercle gives the most extensive attachment and is continued inferiorly as *a stout ridge* much roughened, especially above and below, superiorly for attachment of one tendon of postea spinatus, inferiorly it presents a tubercular termination roughened and curved slightly backwards where teres externus is attached. Just behind the ridge is a triangular roughened space extending to the fourth tubercle. To the anterior part of this the inferior extremity of scapulo-humeralis externus is attached, while its inferior margin marks out the limit of humeralis externus and the line of attachment of caput medium of triceps extensor brachii. The *fourth tubercle*, situated externally to the head, is continuous with the antero-external tubercle. It has the articulatory surface for the scapula extending on to its internal surface, while its external surface affords attachment to one head of postea spinatus, over which the other head plays, with a bursa intervening, to become inserted into the upper part of the ridge. Just below the fourth tubercle is a row of articulatory foramina. The **shaft** of this bone presents four surfaces; the *internal* is straight; superiorly, just below the roughened surface for attachment of subscapularis, the upper head of coracohumeralis becomes attached. Centrally, at the superior part of the middle third, elongated from above downwards, is a roughened ovoid space for insertion of the common tendon of teres internus and latissimus dorsi. Below this, inclined to the anterior surface, is the attachment of the lower head of coraco-humeralis; between the two heads runs a branch of the humeral artery with its satellite vein and nerve to flexor brachii, from behind forwards, but the attachment of teres internus is a little posteriorly placed to their connecting line. Straight down the limb, but in consequence of the position of the bone, obliquely over its inner surface, the humeral artery runs from the inner surface of the shoulder-joint downwards to the inner part of the elbow-joint. At the superior part of the inferior third of the bone it gives off the medullary artery, which pierces the *medullary foramen* here situated. The corresponding vein is more posteriorly placed. The *anterior surface* is superiorly smooth, mainly occupied by flexor brachii running downwards, slightly inclined inwards, after passing over the tubercles of the humerus. The roughened attachment of the long head of coraco-humeralis encroaches upon this surface at the superior part of the inferior third, and is separated by a prominent line running downwards from the ridge on the external surface, from a deep depression just above the lower articulatory surface which almost (in some species of animals quite) extends through the bone as far as the supra-condyloid fossa. This is the *anterior fossa of the humerus.* It presents many foramina, and affords attachment to levator humeri and pectoralis transversus. Externally placed to this the surface at the inferior part is very rough, bounded posteriorly by a ridge which runs upwards and backwards from the outer condyle. This ridge is most prominent inferiorly, and there affords attachment to extensor pedis at its lowest part, while superiorly, and to the roughened surface in front of it, is attached extensor metacarpi magnus; from the inferior part to the inferior prominent extremity of the humeral ridge runs *a band of mixed white and elastic fibres* which gives attachment to panniculus carnosus, levator humeri, and pectoralis transversus, and seems to bind down humeralis externus in the twisted, wide, smooth groove which we may trace from this around the *external surface* of the bone bounded superiorly by the head and the line running from this to the ridge, inferiorly by a much less prominent roughened line winding round from the attachment of subscapularis at the inner surface of the superior extremity as far down as the external part of the outer condyle. The lower part of the *posterior surface* presents **two ridges** running upwards from the condyles of which the inner is largest at its inferior extremity and runs straight upwards, while the outer, at first insignificant, proceeds obliquely inwards and upwards, becoming imperceptibly blended with the middle third of the bone. It was the external margin of this ridge we saw on the outer side of the bone. Between the ridges is the *supra-condyloid fossa*, deepest inferiorly, which serves to accommodate the hamular process of the ulna, and in the fresh subject contains some fat. Inferiorly it presents an articulatory surface concave from side to side continuous with that of the condyles. To the upper part of the **outer ridge** anconeus is attached, while caput parvum arises from the extreme superior part of the inner ridge. From the inferior extremity of the outer, flexor metacarpi externus arises, while the **internal ridge** inferiorly gives attachment to the common originating tendon of perforans and perforatus, and externally is rough for attachment of one head of flexor metacarpi medius and flexor metacarpi internus. The rest of the *inferior extremity* of the bone is occupied by the articulatory surface of the elbow-joint. It consists of two parts or **condyles.** The *inner condyle* is much larger than the outer, and presents one prominent convex surface, decreasing in size towards the *outer condyle* which consists of two convexities. These condyles are connected together by a slight articulatory concavity, which extends posteriorly into the supra-condyloid fossa. On either side of this extremity is a depression for attachment of a lateral ligament of the elbow-joint. In the centre of the articulatory surface is a peculiar roughened depression with rounded edges; this must not be mistaken for ulceration.

RADIUS AND ULNA.

The basement structure of the forearm comprises two bones, both of which, in the ox and other animals, extend to the knee. In the horse, however, the posterior bone or ulna terminates about opposite the superior part of the inferior third of the radius or anterior bone, though traces of it may sometimes be seen united to the inferior extremity at the postero-external part. The **RADIUS** is a long round bone, extending from the elbow to the knee, presenting a shaft, a superior and an inferior extremity. The *shaft* centrally is convex from side to side, and also from above downwards. It is smooth *anteriorly*, and over it play extensor metacarpi and extensor pedis. From the outer margin of the middle third extensor metacarpi obliquus runs obliquely over this surface downwards and inwards. The *posterior surface* is slightly concave from side to side, and to its outer margin the inferior part of the ulna is attached to an extremely elongated triangular roughened surface. The rest of the central part of this surface affords attachment to radialis accessorius and the superior suspensory ligament. The *external* and *internal sharp margins* afford attachment to the faschia which sustains the muscles of the forearm, while more externally placed than the ulna, extensor suffraginis is situated, enclosed between two layers of faschia, one of which is attached in front, the other behind it. The *upper extremity of the radius*, at the extreme superior part, presents an articulatory surface corresponding to that of the inferior extremity of the humerus. It consists of an inner and an outer part, almost separated by a roughened groove, extending forward from the posterior margin of the bone to within about one quarter of an inch of the anterior margin, which is here slightly indented backwards. The *inner surface* for the large internal condyle is concave in all directions, larger than the outer surface, and posteriorly continuous with a synovial facet on the posterior surface of the bone, which is in apposition with the inner synovial facet of the ulna. The *outer surface* is divided into two parts; the inner part is largest in every way, and is continuous inwards at the anterior part with the outer half of the articular surface; outwardly it is continuous with the outer small division. The posterior margin of this surface is continuous with the outer facet, which is on the posterior surface, and articulates with the ulna. The outer circumferent margin of the whole articulatory surface is prominent for attachment of the capsular ligament of the elbow-joint, and both internally and externally below it is a roughened surface (of which the outer is largest) for attachment of the lateral ligaments; from the outer runs extensor suffraginis, while the inner is separated by a smooth portion of bone with slight transverse grooves, from a roughness on the shaft, to which also the inner lateral ligament is attached, while the tendon of humeralis externus runs under this prolongation of the ligament over the smooth surface to become attached to a roughened space on the posterior surface, which extends as far as the inner margin of the union with the ulna. The anterior surface of this extremity presents a prominent roughened **tubercle** continuous with the roughened space on the inner part for attachment of flexor brachii. Externally to this is a smooth surface with a few foramina. On the posterior part of the bone, inclined to the outer side below the synovial facets for the ulna, is a roughened space, to which the fibrous band between the bones is attached spreading beyond the line of the ulna, on either side, in its attachment to the radius. Below this is the radio-ulnar arch formed by the approximation of smooth parts of the two bones, through which a branch of the radial artery (vein and nerve) passes to flexor metacarpi externus from within outwards. This space presents below the *medullary foramen of the radius*, for the artery during its passage gives off the nutritive vessel of that bone. Below this the union of the bones is ossific, except in the young animal, in which it is fibrous. The *inferior extremity* of the radius articulates with the upper row of bones of the knee, and for this purpose presents three continuous articular facets. The *inner* is the largest in every respect. Viewed from behind forwards it is posteriorly convex, anteriorly concave. It posteriorly slightly extends on to the back of the bone. It articulates with os scaphoides. Its anterior, posterior, and inner margins afford attachment to the radio-carpal capsular ligament. Its outer abruptly bends upwards posteriorly in commencing the *articulatory surface for os lunare or middle surface*, which, after running upwards for a short distance at the posterior part, bends outwards, thus producing a convexity posteriorly, a concavity anteriorly, separated from the anterior concavity of the scaphoid surface by a slightly prominent articular ridge. It decreases in size from its inner to its *outer* margin, and is there continuous with the small convex articulatory surface which extends on to the posterior surface of the bone, and comes anteriorly in contact with the cuneiform bone posteriorly with trapezium. The posterior surface of the bone above this articulatory portion presents a roughened prominence, separated from the middle articulatory facet by a deep fossa. This affords attachment to the thick posterior annular ligament of the knee. Each side presents a prominence, the inner is the largest, for attachment of a lateral ligament of the knee. The outer prominence is divided into two parts by a groove running from above downwards, through which extensor suffraginis plays, bound down by the annular ligament. The anterior surface inferiorly presents four prominences for attachment of the annular ligament, and *three intervening grooves*. The *outer groove* gives passage to extensor pedis tendon, the central to extensor metacarpi magnus, the *internal* runs obliquely inwards and downwards as far as the articular surface, and gives passage to extensor metacarpi obliquus.

The **ULNA** is attached (as above described) to the postero-external part of the radius, superiorly by two small synovial articulatory facets connected with the synovial surface of the elbow-joint. Below this by a fibrous articulation, and at the lowest part by an ossific union, and between the ossific and the fibrous portion is the radio-ulnar arch. The ulna is divided into a free and an attached portion. The **attached portion** is prismatic, largest superiorly, and from this tapers downwards, terminating insensibly, blending with the outer edge of the posterior surface of the radius at the superior part of the inferior third. The *outer surface* affords attachment to extensor suffraginis, and slightly to extensor pedis. The *inner* at the radio-ulnar arch to humeralis externus. Its *anterior surface* is that attached to the radius, while its *external* and *internal surfaces* meet in forming a rounded border posteriorly. The **free portion or olecranon** presents two surfaces, three borders. The *anterior border* is articulatory, concave from above downwards, convex from within outwards. It extends most on to the outer surface. Inferiorly it widens, terminating externally and internally in the synovial surfaces of the attached portion of the bone. Frequently only the superior part of this border is smooth, roughness extending completely across it at the inferior part. This surface articulates with the supra-condyloid portion of the articulatory surface of the humerus, thus assisting to form the elbow-joint. Superiorly it forms an acute angle with the *antero-superior margin* which is sharp, and extends obliquely upwards and backwards. This angle is termed the **hook-like or hamular process**. The antero-superior margin posteriorly widens in joining the tuberous superior margin, which anteriorly presents a smooth surface where rests a bursa, and the tendon of triceps extensor brachii playing over this becomes attached to the roughened part. The *external surface* is convex, and roughened along its central line from above downwards for attachment of some of the muscular fibres of caput medium, and more anteriorly of anconeus. Inferiorly it is smooth on becoming continuous with the external surface of the attached portion of the bone. The *inner surface* is smooth and concave. At its junction with the superior margin it affords attachment to one head of flexor metacarpi medius and to scapulo-ulnaris. It blends with the external surface in forming the rounded *posterior margin*, which is continuous with the corresponding part of the attached portion, and which, at its junction with the superior margin, affords attachment to ulnaris accessorius.

The **CARPUS**—" wrist" of human subject, "knee" of horse—

consists of seven or eight small bones arranged in two rows between the radius and metacarpus. The bones composing the two rows are connected together by inter-osseous ligaments, and by other connecting ligaments. The **upper row** is formed of four bones, which form the radio-carpal articulation with the radius; between this and the **lower row**, which is composed of three, sometimes four, bones is the carpal ginglymoid or hinge-like joint, while the lower row forms with the metacarpus an arthrodial or gliding joint, carpo-metacarpal. The *four bones of the upper row* are the scaphoid, lunar, cuneiform, and trapezium; *those of the lower row*, trapezoides, magnum, and unciforme, and sometimes the pisiform bone.

OS SCAPHOIDES is irregularly cubical, presenting six surfaces. It is the inner bone of the upper row and articulates with radius, lunare, trapezoides, and magnum. Both its *inferior* and *superior surfaces* are articulatory, presenting anteriorly a convexity, continued anteriorly on to the outer surface, posteriorly a concavity. But the concavity is most marked on the inferior surface, and while that of the superior surface is continued for a short distance on to the posterior surface of the bone, that of the inferior surface does not completely cover that surface, which posteriorly presents a roughened part, projecting inwards. The *upper surface* articulates with the inner prominent portion of the radius. The inferior surface anteriorly (the convex part) with os magnum, posteriorly (the concave part) with os trapezoides. The space between the reflections of these synovial surfaces which articulate with os lunare into the *outer surface* of the bone is flat but rough for attachment of inter-osseous ligaments, as also is a concavity behind it, separated into a deep superior part, and an inferior part by a sharp ridge. Posteriorly the concavity is bounded by a roughened ridge, a continuation of the *posterior roughened surface* of the bone, which is tuberous below and superiorly presents a continuation of the superior articular surface. This surface is continuous inwardly with the *interior* and *anterior surfaces*, both of which are roughened for attachment of connecting ligaments superiorly and inferiorly, and of the annular ligament centrally.

OS LUNARE is the centre bone of the upper row. It is elongated from before backwards, and widest superiorly. It articulates with the radius, scaphoides, cuneiforme, magnum, and unciforme. It presents six surfaces. The *superior surface* is articulatory, convex anteriorly, concave posteriorly, terminating behind in a point which extends upwards. Its outer margin also anteriorly terminates in a point. This surface articulates with the central part of the inferior articulatory surface of the radius, the vertical portion of which corresponds to a continuation of this smooth articulatory surface on to the inner surface of os lunare, which surface, at its antero-superior part, also comes in apposition with os scaphoides. This surface, too, is continued on to the external part of the bone anteriorly, where it plays on os cuneiforme. The *inferior surface* presents two parts, inner of which is convex anteriorly, concave posteriorly, extending from the anterior to the posterior margin of the bone. It articulates with os magnum, and anteriorly is continued on to the internal surface, forming a small articulatory facet for os scaphoides. The outer portion of the inferior surface is convex, anteriorly placed and comes into connection with os unciforme. It is continuous with the lower facet on the outer surface, on which os cuneiforme plays. The *anterior surface* is slightly roughened for attachment of connecting and the annular ligament. The *posterior surface* superiorly is concave, inferiorly prominent; it also gives attachment to the annular ligament. The *outer surface* presents posteriorly, the outer surface of this roughened prominence antero-superiorly, and antero-inferiorly are the articulatory facets for os cuneiforme. The space between them is concave and rough for attachment of interosseous ligaments. The *inner surface* superiorly along its whole length presents the vertical articulation for the radius, which anteriorly is continuous with a small surface for os scaphoides, which is also in contact with a corresponding facet at the antero-inferior part continued from the inferior surface. The space between these is rough for attachment of interosseous ligaments.

OS CUNEIFORME is situated at the outer part of the upper row. It articulates with radius, lunare, unciforme, and trapezium. It presents two surfaces and four borders. The *external surface* is convex in every direction, and is roughened for attachment of the annular, connecting, and outer lateral ligaments. Along its postero-superior margin it presents a roughened groove for attachment of the capsular ligament of its articulation with trapezium. The *internal surface* presents anteriorly two smooth articulatory facets, one superiorly continued as far as the anterior margin, continuous with the articulatory surface on its superior margin; one inferiorly, separated by a roughened space from the anterior margin, continuous with the articulatory surface of the inferior margin. Between and behind these this surface is roughened for attachment of interosseous ligaments, posteriorly it terminates in a point. The *superior margin* presents a smooth triangular concave surface for articulation with the radius continuous with the superior one on the internal surface for lunare. The *inferior surface or margin* presents an articulatory surface with the unciform, which anteriorly is concave, posteriorly at the external side convex. It is continuous with the inferior facet on the inner surface for lunare. It is separated by a roughened space from the *posterior margin*, which presents a concave ovoid articulatory surface extending obliquely forwards and upwards; with this the inferior articulatory surface of trapezium comes in contact. It is separated from the inferior articulation by a roughened depression, to which its capsular ligament is attached. The *anterior margin* of the bone is also roughened for ligamentous attachment.

OS TRAPEZIUM is situated at the postero-external part of the upper row. It is flattened from side to side, and slightly curved inwards. It presents two surfaces and four margins. The *anterior margin* has two articulatory facets, the superior concave for articulation with the radius, and extending on to the internal surface of the bone. It is separated by a deep perforated groove from the inferior facet, which is below and slightly externally placed, convex for articulation with cuneiforme; internally placed to this the anterior margin is rough. The *superior* and *inferior margins* are rough, and extend from the anterior to the posterior margin. The superior is sharpest, and from about its centre over the external surface of the bone a roughened groove extends towards the infero-anterior articulatory facet. Flexores metacarpi externus et medius are attached to the upper margin, and through the groove on the external surface a rounded tendon of flexor metacarpi externus runs to the outer small metacarpal bone. This groove presents a prominent roughened edge, separating it from a depression; the rest of the *external surface* is convex and roughened for attachment of the annular ligament, which is also attached to the inferior margin and to the *posterior margin*, which is rounded and rough. The *internal surface* is concave and slightly rough for attachment of the annular ligament. At its antero-superior part it presents a small articulatory facet continuous with the superior facet on the anterior margin. From this a ridge runs downwards and backwards obliquely to the inferior margin.

OS TRAPEZOIDES is the inner bone of the lower row, and articulates with magnum, scaphoides, metacarpi magnum, and parvum internum (sometimes with pisiform). It presents three surfaces. The *supero-internal surface* presents superiorly a rounded convexity, which commencing in a point at the outer margin anteriorly occupies the whole upper part of the bone, and passes to the postero-internal part, extending about halfway down to the lower surface. Around it runs a narrow groove, except at the outer part, where it is continuous with a small elongated facet, articulatory for os magnum. The rest of this surface is roughened for attachment of the annular, inner lateral, and connecting ligaments. Along the inferior margin is a groove which bounds the inferior articulatory surface. The *inferior surface* is mainly occupied by a flat facet for articulation with the inner small metacarpal bone, elongated from before backwards, posteriorly continuous with a smaller facet for os metacarpi magnum, and this in turn is continuous with a still smaller facet for os magnum. This articulatory surface is bounded on every side by a groove, except on the outer, where it is reflected on to the outer surface for articulation with os magnum. Between the superior facet (continuous with the upper surface) and the inferior facet (continuous with the lower surface) the *outer part* of the bone is roughened for attachment of interosseous ligaments. Postero-inferiorly is a peculiar prominence presenting a sharp point inwards, bounded by the posterior surface for os magnum posteriorly, inferiorly by the facet for os metacarpi magnum.

OS MAGNUM is the central bone of the lower row, and articulates with lunare, scaphoides, trapezoides, unciforme, metacarpi magnum, et parvum internum. It has five surfaces, which give it a triangular appearance, and is flattened from above downwards, having a base or anterior surface, and an apex or posterior angle which is rounded. The *superior surface* is triangular and presents two articulatory portions which blend anteriorly. The outer consists of two triangular portions united at their apices, occupying the inner half of the surface, extending from the anterior margin to the posterior angle. The posterior triangle, convex and rounded, extends on to the posterior angle of the bone. The anterior triangle is continuous on its outer margin with a facet, on the anterior part of the outer surface for os unciforme. Internally it is continuous with the inner articulatory part of this surface, which is triangular, occupying the inner angle of the surface, separated postero-externally from the other part by a perforated concavity. It is slightly continued on to the inner surface of the bone for articulation with trapezoides. With this part os scaphoides comes in contact, while os lunare occupies the outer part. The *inferior surface* is flattened, slightly convex anteriorly, slightly concave posteriorly, and has four margins. The anterior margin is convex, extending farthest backward on the inner side. The posterior is not half the length of this, and is perfectly straight. The outer margin is straight, but centrally presents a deep depression, roughened, and extending towards the centre of the surface. Along the inner margin, at its anterior part, is a small facet for articulation with the inner small metacarpal bone; this is continuous with the main surface for os metacarpi magnum, and is continued on to the inner surface, forming a small facet for the trapezoid. The inner surface presents three facets for trapezoides—one antero-superiorly elongated, one antero-inferiorly, shorter but deeper, and one posteriorly extending from the posterior part of the inner concavity of the inferior surface; this looks forwards. The rest of the surface presents irregular roughened concavities and convexities for attachment of interosseous ligaments. The *outer surface* presents articulatory facets for union with os unciform, between which the bone is rough for attachment of interosseous ligaments. The *anterior surface or base* forms an irregular parallelogram extended from side to side, which centrally presents a roughened prominence for attachment of the annular ligaments, superiorly and inferiorly a groove bounded by the articular facets of the upper and lower surfaces.

OS UNCIFORME.—The outer bone of the lower row articulates with cuneiforme, lunare, magnum, metacarpi magnum, et parvum externum. It presents four surfaces, and a posterior very prominent angle. The *supero-external* *surface* presents an articulatory convexity extending obliquely downwards, outwards, and backwards. The outer part of this articulates with cuneiforme, the inner with lunare and extends to a superior facet on the inner surface for articulation with magnum. The *inferior surface* presents three articular parts, the anterior, extending obliquely outwards and backwards, internally continuous with the inferior facet for magnum on the internal surface. It articulates with metacarpi magnum, as also does the posterior part, which slightly meets the former by its anterior angle, but is separated from it inwardly by an indentation on its inner margin, externally by the middle part of the surface, which articulates with the outer small metacarpal bone. This surface presents a semicircular figure, and a groove extends round its outer margin. The *inner surface* so exactly corresponds with the outer surface of os magnum that one description will suffice for both. The posterior facet is on the inner side of the process, the outer, posterior, and superior parts of which are rough, as also is the *anterior surface*, affording attachment to the annular ligaments, bounded on every side by grooves mapping out the articulatory facets of the neighbouring surfaces.

OS PISIFORME is a small osseous nodule found on the inner side of the lower row, articulating with trapezoides, and the inner small metacarpal bone, for which it presents small articulatory facets. Its presence is by no means constant; it is relatively larger in the ass than in the horse.

THE METACARPUS

is formed of three bones, the largest of which is centrally placed, with a smaller one on each side of it. It extends from the knee to the fetlock, but the small bones only proceed downwards part of this distance.

OS METACARPI MAGNUM (CANON OR SHANK BONE) is a long, round bone, extending from the knee to the fetlock, having two extremities and a shaft. The **shaft** is flattened from behind forwards, and presents two surfaces connected by two rounded margins. The *anterior surface* extends in a straight line from above downwards, is convex and rounded from side to side, becoming flattened, and more expanded superiorly and inferiorly. It is for the most part smooth, and is separated from the skin only by the expanded tendons of extensor pedis and extensor suffraginis, which run from above downwards inclined to the outer side, and are closely connected together. At its extreme superior part it presents a groove marking out the limit of the superior articulatory surface, and below this is roughened, especially on the inner side, where is a roughened prominence for attachment of extensor metacarpi magnus. The *posterior surface* is flattened, and on either side presents a roughened surface for attachment of the metacarpal interosseous ligaments, binding the small to the large bone. These surfaces are widest superiorly and inferiorly, taper to a point slightly converging, extending to about the inferior part of the middle third of the bone. The space between them is concave, and at the inferior part of the superior third inclined to the inner side presents the *medullary foramen*, through which a branch of the small metacarpal artery runs to supply the bone. The extreme superior part of this surface affords attachment to the superior extremity of the superior sesamoideal ligament, the anterior surface of which lies in contact with the greater part of the rest of this surface, which, superiorly convex, at the lower part of the middle third becomes concave centrally, again becoming prominent, especially along the middle line in reaching the inferior articulatory surface. Here it is pierced by a row of articular foramina. The *margins* of the bone superiorly are roughened for attachment of the capsular (carpo-metacarpal) and annular ligaments of the knee, centrally are smooth, forming with the small metacarpals grooves running from above downwards. Inferiorly they are rough for attachment of a ligament from the inferior extremity of the small metacarpal bone, and one from the lateral part of the sesamoid bone, and below this present each side a rounded concavity for attachment of the lateral ligaments of the fetlock. The *upper surface* of the bone presents four distinct articulatory facets. One is by far the largest and slants obliquely inwards, and occupies the inner two-thirds of the surface. It articulates with os magnum and presents four margins. The anterior margin has two convexities and extends farthest backwards on the inner side, and there meets the inner margin which runs obliquely from within outwards and backwards. It is divided into an anterior and a posterior part by a break in the margin. The anterior part is continued on to the posterior surface in forming the small anterior articulatory facet for the inner small metacarpal bone. The posterior part is continuous with a small facet for os trapezoides which, in its turn, joins an extremely small one for os metacarpi parvum internum. The posterior margin is perfectly straight and about half the size of the anterior. The outer margin anteriorly and posteriorly is continuous with circulatory facets sloping outwards, separated by a deep roughened depression. These articulate with os unciforme, and are each continuous with an articulatory facet on the posterior surface for the external small metacarpal bone. The *inferior articulatory surface* is convex in every direction, and consists of three prominences running from behind forwards. The central prominence is most marked, narrowest, and extends farthest both anteriorly and posteriorly. It is wholly articulatory. The inner and outer prominences consist of surfaces sloping towards the central ridge. They are coated with cartilage continuous with that of the central ridge. In most cases the inner is larger than the outer articulatory surfaces.

OSSA METACARPI PARVA are extremely elongated small bones, placed at the lateral, inclined to the posterior, part of the large metacarpal bone. They decrease in size from above downwards, terminating *inferiorly* opposite the inferior part of the middle third of the large bone in small rounded bulbs, from which the superior ligament of the pad, and a small ligament to the infero-lateral part of the large bone, run downwards. The *superior extremity* is rough and tuberous for the most part, while superiorly it is more or less occupied by articulatory surfaces, which extend slightly on to the attached part as two synovial facets for os metacarpi magnum. Below this the whole *attached surface* is roughened for the white fibrous connection with the large bone. The *inner* and *outer surfaces* are smooth, and from the inner runs the anterior lumbricus muscle, while the outer, with the large bone, forms a smooth groove.

The *inner small metacarpal bone* at its upper extremity is for the most part flat for articulation with os trapezoides. Its outer margin anteriorly presents an almost vertical articulatory facet to blend with that of the large bone, which articulates with os magnum, while postero-externally is a small facet for connection with os pisiforme. It gives attachment to extensor metacarpi obliquus and flexores metacarpi internus and medius, the latter through the medium of the annular ligament of the knee. To it also is attached the inner lateral ligament of that joint. It is larger and longer than the outer small bone.

The *outer small metacarpal bone* supero-externally is rough for attachment of the outer lateral, and the annular ligament of the knee, also of flexor metacarpi externus. Supero-internally it presents an articulatory surface, flat and slanting inwards, on which os unciforme glides.

SESAMOID BONES, two in number in each limb, are situated at the fetlock behind the inferior extremity of the large metacarpal bone on which they play, for their anterior surfaces are articulatory, and they are so united together that the surfaces of the two bones form the counterpart of the posterior part of the inferior extremity of the large metacarpal bone. Each bone presents four surfaces. The *anterior surface* is triangular and articulatory, consists externally of a large smooth surface, which inwardly becomes reflected backwards in forming with its fellow the counterpart of the central ridge of the inferior extremity of the large metacarpal bone. This surface inwardly is connected by the *inner margin* to the *posterior surface*. This margin gives attachment to the intersesamoid ligament which binds the bones together, and thus renders their posterior surfaces which are coated with fibro-cartilage, a convenient groove for the flexor tendons to play over. The *inferior surface* is rough, triangular, and presents foramina; from it the inferior sesamoid ligaments extend. The *outer surface* inclines obliquely upwards and inwards in such a manner as to form a point superiorly in meeting the anterior and posterior surfaces. The posterior margin of this surface presents a prominent ridge, and anteriorly some large foramina. From the ridge the lateral sessamoid ligaments run, and also to the ridge are attached those portions of the superior sesamoideal ligament which pass to extensor pedis, and also the annular ligaments of the fetlock.

OS SUFFRAGINIS is the long round bone which extends from the fetlock to the pastern joint. Its *shaft* is short and slightly compressed from before backwards. The *anterior* *surface* is convex from side to side, and is mainly occupied by the extensor pedis tendon with the bands running to it from the superior-sessamoideal ligament. It is, therefore, smooth for the most part, but above and below it becomes roughened. Superiorly it is prominent for attachment of the tendon of extensor suffraginis (or ligamentum extensorum). This prominence is slightly inclined to the outer side. The *posterior surface* is also convex, and superiorly presents externally and internally two prominent angles from which two ridges converge to form a prominent rough surface at the central part, to which the short part of the inferior sessamoideal ligaments is attached. To the angles, are attached white fibrous bands, which converge in becoming attached to the posterior surface of the perforatus tendon; they are parts of the annular ligament of the fetlock. The *margins* of this bone lie in contact with the plantar vessels and nerves, and inferiorly present a roughened line, to which the superior ligament of the navicular bone is attached in blending with the lateral ligament of the pastern joint. The *superior articulatory surface* of the bone is in every respect concave, and presents three concavities, the inner and outer of which are triangular; the inner is slightly the largest. The central concavity is deepest and narrowest, extending from the anterior to the posterior surface, while between it and the above-mentioned ridges is a rougher space perforated by numerous foramina. The *inferior extremity* is much smaller than the superior, and presents an irregularly heart-shaped articulation, which, by uniting with the upper extremity of os coronæ, forms the pastern joint. This is obtusely rounded anteriorly, extending slightly on to the anterior surface, and consists of two rounded convexities, which meet forming a shallow groove. They extend considerably on to the posterior surface, and are here separated by a roughened indentation; the inner is slightly the largest. On either side, slightly inclined to the anterior part, just above the articulation, the lateral ligaments of the pastern are attached. Posteriorly, above the articulatory surface, are numerous foramina.

Ossa coronæ, pedis, and naviculare will be examined in dissection of the foot.

BONES OF THE HIND LIMB.

The posterior cavity of the body is termed the pelvis. It is most marked in the skeleton for in the fresh subject it is anteriorly confounded with the posterior part of the abdomen. Its basement structure consists of two bones (each composed of smaller portions connected together by ligaments), superiorly is the sacrum which has been described as part of the spinal column. Inferiorly and laterally are the ossa innominata.

OSSA INNOMINATA are two in number, one on each side, and these are united at the inferior part centrally forming the *symphysis*. These bones originally consist each of three parts, but in the adult these parts are firmly united to form one bone. Supero-anteriorly is the *ilium* forming the lateral boundary of the pelvis at the anterior part; antero-inferiorly is *os pubis;* postero-inferiorly *os ischium*. The three unite in forming a cotyloid cavity termed the **acetabulum**, coated with articular cartilage with a gap extending from its inner and anterior part towards the centre, where ligamentum teres which serves to maintain the cavity in contact with the head of the femur, is attached. It is deep in all parts and is rendered more so by a band of fibro-cartilage (circumferential or cotyloid) which is attached around its margin, and is continued across the gap forming the transverse ligament, and binding down the pubio-femoral ligament in its passage to the head of the femur. The ischium forms the major portion of this cavity (about three-fifths), while the pubis forms the least, a little less than one-fifth, os ilium forming a little more than one-fifth; but a considerable part of the pubis is occupied by the roughened groove extending inwards. The acetabulum looks outwards and slightly downwards and backwards.

OS ILIUM is a flat bone, triangular in shape, placed obliquely from above downwards and outwards. It is triangular, and presents two surfaces, three margins, and three angles. The *antero-inferior angle* is termed the **antero-inferior spinous process**, it is rough and tuberous, elongated from above downwards presenting four prominences, two superiorly, two inferiorly. It affords attachment internally to iliacus, longissimus dorsi, obliquus abdominis externus, obliquus abdominis internus, transversalis abdominis, and quadratus lumborum. Externally gluteus maximus is attached to it, and from its inferior tubercles runs the common tendon of tensor vaginæ and gluteus externus. The *superior angle* is the **supero-posterior spinous process**, it is elongated, flattened from within outwards, and superiorly is curled slightly outwards. To it runs the superior ilio-sacral ligament, from its posterior margin the inferior ilio-sacral ligament. To its inner surface is attached longissimus dorsi, and to its external surface gluteus maximus; while gluteus externus, in some cases, seems to arise from its summit. The *inferior angle* becomes rounded in forming the neck, which is convex, and below dilates to form the anterior part of the acetabulum inferiorly, the union with the neck of the ischium, externo-superiorly, the junction with the os pubis antero-internally. The *external surface* of the ilium is called the **dorsum**, and is concave, looking upwards, outwards, and backwards. It becomes convex in forming the **neck**. Superiorly to the upper margin it affords attachment to gluteus maximus, and this muscle is firmly attached to it as far as a crescentic groove, running across its neck where gluteus internus commences, and becomes attached as far down as the acetabulum. The *margin connecting the anterior and superior angles* is slightly concave, and is rough and thickened. It is termed the **crista**, and affords firm attachment to the tendinous structure of longissimus dorsi. The *anterior margin* is concave, and looks downwards and outwards, it is smooth, affords internally attachment to iliacus, externally to the glutei muscles, and inferiorly at the neck becomes widened, forming two roughened fossæ separated by a smooth space. To these, one of which is inclined to the dorsum, the other to the venter surface, are attached the two heads of rectus femoris; between them is a small quantity of fat. Below this ilio-femoralis arises, above it psoas parvus is inserted. The *posterior margin* is divided into three parts. The superior third gives attachment to the inferior ilio-sacral ligament, the middle is smooth for passage of gluteal vessels and nerves round it from within outwards, the inferior forms a very sharp prominent ridge continued on to the neck of the ischium, to which the sacro-sciatic ligament is attached. This is the **ischiatic spine**, and externally is continuous with a small but very rough space for attachment of part of gluteus internus. The *internal surface* of the ilium is termed the **venter**. It is convex for the most part, and is divided into two portions by a roughened articulatory surface for the transverse process of the first sacral bone. This articular surface (*auricular facet*) looks obliquely downwards and forwards, and is smoothest centrally, where is sometimes found a small synovial cavity. Its outer margin is very rough for attachment of the sacro-iliac ligament. The surface anteriorly placed to this, which helps to form the brim of the pelvis, is triangular, terminating in a point inferiorly. It gives attachment to the iliacus, and that portion of it which forms the anterior part of the neck presents transverse grooves, through which runs the arteria innominata from behind forwards in its course to the triceps cruralis. This surface is separated from the posterior surface, which is much narrower, and gives origin to pyriformis, by grooves running directly downwards to the obturator foramen, through which the obturator vessels and nerve pass. In contact with this surface superiorly and under the first sacral transverse processes are the iliac arteries and their commencing branches, also the external, internal, and common iliac veins.

OS ISCHIUM occupies the postero-inferior part of the pelvis, and consists of three processes with the body at which they join. The *antero-external process* runs forwards to help to form the acetabulum, consequently the superior part of its anterior surface is roughened for articulation with the ilium and its infero-anterior part for union with os pubis. It is smooth, both externally and internally; externally, in forming the **neck of the ischium**, over which obturator internus and pyriformis play, the latter muscle is attached to a roughened line on the upper part of the process; internally in forming with the *antero-internal process* an arch with rounded margin, which looks forwards and gives attachment to obturator externus, and assists in forming the obturator foramen. The anterior extremity of the last-mentioned process joins os pubis, while its inner margin, meeting its fellow, assists the corresponding part of the body in forming **symphysis ischii**, the union becoming strengthened below by a connecting ligament, particularly thick opposite the posterior margin of obturator foramen, in becoming reflected down to form *the triangular ligament of the penis* of the male. The superior margin of symphysis pubis et ischii affords attachment along almost its whole length to a reflection inwards of the sacro-sciatic ligament, which covers the origin of obturator internus from the superior part of the margins of the foramen of the same name, separating it from the pelvic viscera. The *posterior process* of the ischium is triangular and tuberous at its extremity, which extends outwards and backwards, and is termed the **tuberosity of the ischium.** It presents a small rough ridge, extending forwards along its outer margin to the neck, from which ischio-femoralis arises. To the posterior tuberous part triceps abductor femoris, biceps rotator tibialis, ischio tibialis and erector penis are attached. The inner margin of this process slopes inwards towards its fellow, which it joins at an oblique angle at the symphysis. This is the ischial arch, and round it the urethra passes from the pelvis into the penis. In so doing it is bound down by a small ligament and by the crura, which, being attached on either side, converge to form corpus cavernosum.

OS PUBIS consists of three sides and three angles; its *inner margin* lies in contact with its fellow, and thus forms **symphysis pubis**, continuous with and analogous to symphysis ischii. The *posterior angle* is united to the antero-internal process of the ischium, and from it the outer margin forms an arch running forwards and outwards, which, after completing the *obturator foramen*, terminates in the junction with os ischium and os ilium, forming the acetabulum. The *anterior margin*, forming the brim of the floor of the pelvis, is at first smooth, centrally presents a rough prominence, **pectinean tubercle**, to which pectineus is attached, and inwardly these two pectinean tubercles are connected by a dense fibrous band, which affords attachment to the abdominal muscles anteriorly, and posteriorly is partly continuous with the connecting ligament of the symphysis, while the rest of it passes through a groove, which occupies the under surface of os pubis, extending outwards to the pubic notch of the acetabulum. Symphysis pubis superiorly forms a smooth concave surface for the bladder to rest upon, and externally to this the inward reflection of the sacro-sciatic ligament is attached. To the under surface of the whole symphysis the common tendon of the graciles is attached centrally, and on either side a biceps adductor femoris.

The **FEMUR** is a long, round bone, extending from the hip to the stifle. It is the largest bone in the body, and centrally is rounded, but superiorly becomes flattened from behind forwards. It consists of a shaft and two extremities. The **shaft** presents four surfaces. The *anterior surface* is in all parts convex and rounded, though in blending with the upper extremity it is inclined to become flat. It is directly continuous with the inner and outer surfaces. The *inner surface* is rounded. Supero-posteriorly it presents a roughened prominence, elongated from above downwards, continuous by two roughened sharp lines with the head. The anterior of these sharp lines marks out the attachment of vastus internus. This prominence is **trochanter minor internus.** To it are attached psoas magnus and iliacus. Below it the surface is roughened for attachment of pectineus and of one head of biceps adductor femoris; still lower down is an oblique smooth surface, extending from above downwards and backwards for passage of the femoral artery and vein, at the upper part of which, opposite the centre of the bone postero-internally is the *medullary foramen*. Still lower down this surface is very rough for attachment of the other head of biceps adductor femoris. The *posterior surface* of the bone superiorly is very wide, presenting a flat surface between the two small trochanters. Slightly inclined to the outer side of the centre of this part is a roughened surface for attachment of ischio-femoralis and a portion of triceps abductor femoris. From the upper part of trochanter minor externus a thick ridge runs upwards towards the posterior part of trochanter major; this forms the outer boundary of the **trochanteric fossa,** a deep depression on the supero-posterior and external part of the bone, into which are inserted the two obturator muscles and pyriformis, with gemini or ischio-trochanterius. The middle third of this surface is roughened and continuous with that on the inner surface for pectineus and biceps adductor femoris; below this, separated by the smooth oblique groove for the femoral artery, is a triangular space extending as far as the outer condyle for attachment of the lower head of biceps adductor femoris. The *external surface* presents superiorly the ridge running from trochanter major to trochanter minor externus, to which the posterior head of gluteus maximus is attached. This culminates in a long process flattened from before backwards, curled forwards, and tuberous at its free margin, **trochanter minor externus,** to which gluteus externus is attached. Below this for a short distance it is smooth, and still lower down is a deep fossa which looks outwards and backwards, and is roughened for attachment of gastrocnemius internus. It is bounded by ridges, of which the outer is most prominent and roughened for origin of the outer head of gastrocnemius externus; while the other head arises from near the lower insertion of biceps adductor, separated from the former and from internus by the femoral artery. The inner ridge is smooth. The **superior extremity** is divided into two parts—head and trochanter major. The *head* is internally placed, and consists of a rounded articular surface, convex in every direction, with a break in its margin at the internal part, which, running upwards, leads into a deep groove extending to the centre of the head, roughened for attachment of ligamentum teres and the pubio-femoral tendon of rectus abdominis. Around the outer margin of this surface is attached the capsular ligament of the hip-joint. From the head outwards, bounded posteriorly by the trochanteric fossa, anteriorly by the dilated upper extremity of the anterior surface, is a ridge of bone, which externally becomes much expanded in forming an irregularly concave surface, looking inwards, perforated by numerous foramina, to which gluteus internus becomes attached. This is the inner surface of **trochanter major,** which consists of two parts, the *anterior* one rounded and rough, presenting on its external surface centrally a smooth spot for a bursa, below which is a ridge for insertion of the anterior head of gluteus maximus. The *posterior part* is much larger, and superiorly presents a roughened extremity for attachment of another head of gluteus maximus, and from this a roughened ridge forming the outer boundary of the trochanteric fossa runs to trochanter minor externus, to which ridge the third head of gluteus maximus is attached. At the smooth surface between trochanter major and the head ilio-femoralis is attached to a thin roughed line running from above downwards. The inferior limit of trochanter major and the anterior margin of the ridge running to the external small trochanter mark out the superior attachment of vastus externus.

The **inferior extremity** of the femur may be divided into an anterior and a posterior part. The *anterior* consists of two articular *ridges,* which look downwards and forwards, which form a continuous articulatory surface for the patella to play over. The *inner* is much the largest, especially supero-anteriorly, where it becomes bulbous, with an inclination inwards. The *outer* decreases slightly in size posteriorly. The non-articulatory parts of these ridges are rough for attachment of the capsular ligament of the stifle. The *posterior* part of the inferior extremity consists of two *condyles* separated by the **intercondyloid fossa**, which runs from behind forwards. It is roughened for attachment at its postero-internal part of the postero-superior ligament of the outer interarticular disc of cartilage. Just in front of this the *outer* condyle is undermined for attachment of the superior extremity of one crucial ligament, while the *inner* similarly affords attachment to the other. Anteriorly are numerous foramina. The *outer condyle* is the largest, and in front of its inferior extremity is a deep depression, whence arises the common tendon of flexor metatarsi and extensor pedis. Behind and above this is a ridge running backwards, to the anterior part of which the tendon of popliteus is attached, to the posterior parts the outer lateral ligaments of the patella and of the stifle joint. The inner surface of the *internal condyle* is roughened, and presents a prominence to which are attached the corresponding inner lateral ligaments. The articulatory surface of the inner condyle is continuous with that of the inner ridge.

PATELLA is a floating bone, situated at the infero-anterior part of the femur. It is the "knee cap" of the human subject, and has three surfaces and three angles. The *superior surface* looks slightly backwards, it is biconcave (slightly) affords attachment to triceps cruralis, and is bounded anteriorly by a prominent margin, posteriorly by a line, convex externally, concave internally. These meet at a rounded angle outwardly, from which run the outer patellar lateral and straight ligaments, and at an acute angle inwards, where the bone is continued by a fibro-cartilaginous prolongation, from the extremities of which pass the inner lateral and inner straight patellar ligaments. The *anterior surface* is rough and convex, affording attachment outwardly to triceps abductor femoris. Its superior margin is rounded. Its inferior margin comes to a point centrally; just above this is a smooth spot for a bursa and still higher up is roughened for attachment of the central straight ligament. Externally to this is the attachment of the outer straight ligament. The *posterior surface* of the bone is articulatory, consisting of an outer small part connected by a broad prominent ridge with the inner large part, which superiorly is concave and extends downwards, inferiorly inclines forwards, terminating internally in a point. The outer part plays over the outer ridge of the femur, the inner over the inner ridge, the surface being completed by the cartilaginous prolongation inwardly. Around this surface the capsular ligament is attached externally and internally.

The **TIBIA** is a long round bone extending from the stifle to the hock. It articulates superiorly with the femur through the medium of the semilunar discs of cartilage, and directly inferiorly with the astragalus; outwardly with the fibula. It consists of a shaft and two extremities. The **shaft** is prismatic presenting three surfaces, antero-external, antero-internal, and posterior. The *anterior surfaces* centrally blend, forming a rounded margin, but superiorly become expanded and unite in forming a **sharp ridge**, which increases in size as it proceeds upwards. To the internal surface of this ridge the expanded tendinous layer common to ischio-tibialis, sartorius, and gracilis is attached along its whole length, while biceps rotator tibialis, after playing over a bursa on this surface is attached to its prominent edge about the centre. Along the sharp anterior margin externally is attached the aponeurosis of triceps abductor femoris, while to the external surface of the ridge, the free margin of which slightly bends outwards, is attached the muscular portion of flexor metatarsi. The superior extremity of the ridge is triangular, bifurcating in forming two prominences separated by a smooth groove. To the upper part of these prominences the inner and outer straight ligaments of the patella are attached. While the central straight ligament, after playing over a bursa situated in the groove, becomes attached to the point of bifurcation. The inner prominence is much the largest. The *antero-internal surface* is immediately subcutaneous, and over it runs vena saphena from before backwards and upwards. The *antero-external surface* is occupied by the passage of flexor metatarsi and extensor pedis and is smooth. Inferiorly this surface becomes wider on approaching the inferior extremity and internally it presents a small prominence roughened for attachment of the supero-anterior annular ligament of the hock, which runs obliquely downwards and outwards, over the tendons of flexor metatarsi and extensor pedis towards the infero-external part of the bone. The *posterior surface* is divided into two parts. One superiorly placed, bounded superiorly by the superior margin of the bone, internally by the superior half of the inner margin, externally by a line running from without inwards, obliquely downwards. These enclose a triangular space to which popliteus is attached. The rest of this surface is roughened by ridges running obliquely inwards and downwards, to the outer part of which is attached flexor pedis perforans to the inner flexor pedis accessorius; these two muscles are separated by the posterior tibial artery, which runs through a smooth groove, which at the inferior part of the superior third of this surface presents the *medullary foramen*. The *inner margin* of the bone is inferiorly unoccupied. Its superior part affords attachment to popliteus and about the centre of this half is a remarkable prominence serving to mark out the course of the tendon of biceps rotator tibialis. Superiorly this margin gives attachment to the inner lateral ligament of the stifle joint. The *outer margin* superiorly is roughened, affording attachment to the outer lateral ligament; below this is a roughened line to which the upper extremity of the fibula is attached by white fibrous tissue; below this a smooth surface helping to form the fibular arch through which the anterior tibial vessels pass. This margin as far down as the inferior extremity gives attachment to the white fibrous tissue which connects it to the fibula. Its superior part is separated from the anterior ridge of the bone by a deep smooth rounded groove, in which lies a bursa for passage of the tendon common to flexor metatarsi and extensor pedis. Flexor pedis perforans extends as far up as the posterior part of the expanded superior extremity of this margin.

The **superior surface** of the bone is divided into two articular surfaces by a **spine** which runs from behind forwards, presenting an irregular groove extending from the perforated line which separates the anterior ridges from this superior surface, to the posterior margin. It is rough and prominent, anteriorly affording attachment to one of the crucial ligaments, and to the anterior ligaments of the semilunar cartilages. Behind this it is concave and perforated, affording attachment to the posterior ligament of the inner cartilage, while to its junction with the posterior margin are attached the postero-inferior ligament of the outer cartilage and the other crucial ligament. The inner side of the spine is by far the most prominent, and its external surface is articulatory continuous with the *inner articulatory facet*, which is ovoid, bounded by a roughened margin and lying in contact with the under surface of the inner cartilage and with the inner condyle of the femur. The *outer articulatory surface* is triangular, concave, and looks slightly backwards, and is in contact with the under surface of the outer cartilage and with the outer condyle. The **inferior extremity** is quadrilateral and consists of two articular grooves running obliquely from within outwards and forwards, which come in contact with the superior articulatory surface of astragalus. The *inner groove* is the deepest and the narrowest and extends farthest backwards. The *outer* in some cases consists of two portions of bone connected by fibrous tissue, *the outer of them being the inferior extremity of the fibula*. The *outer margin* is rough and prominent, being termed the **outer malleolus**, over which a groove for peroneus runs downwards. It affords attachment to the capsular, outer lateral and supero-anterior annular ligament of the hock. The *posterior margin* is rough for attachment of the capsular ligament, and comes to a point centrally. The *inner margin* is rough, and posteriorly presents a groove running from above downwards for flexor pedis accessorius, anteriorly the **internal malleolus**, larger than the outer, which gives attachment to the inner lateral and capsular ligaments. The *anterior margin* is straight, and rough for attachment of the capsular ligament.

The **FIBULA** is a small very much elongated piece of bone, situated at the external part of the tibia, rounded for the most part and by its inner part connected with the tibia by white fibrous tissue, sometimes continued as far as the true hock joint. Sometimes its middle part degenerates into a white fibrous band. Superiorly it slightly expands in forming a button-like extremity flattened from without inwards, the inner surface of which is roughened for attachment of white fibrous tissue connecting it to the supero-external part of the tibia. Below this the *surface* (*internal*) is smooth, forming (with the tibia) the **fibular arch** through which the anterior tibial vessels pass. The *external surface* of the button-like superior extremity is occupied by the attachment of the outer lateral ligament of the patella. The greater part of the remainder of the external surface gives attachment to peroneus.

TARSUS, HOCK (ANKLE OF HUMAN SUBJECT),

consists of a number of small bones resembling to a certain extent those of the knee. It occupies the space between the tibia and the metatarsus, and consists of *two rows of bones*, of which the *superior* is formed by two somewhat large bones, astragalus and calcis, while the *inferior* consists of four smaller bones, and is split into two rows on its inner side, so that while cuboides occupies the whole depth of the row on the outer side, cuneiforme magnum forms the upper and cuneiforme medium and parvum form the lower subdivision on the inner side. It is between these bones on the inner side that "spavin" generally first manifests itself.

OS CALCIS is the postero-external bone of the upper row, and articulates by three facets anteriorly with astragalus, inferiorly with cuboides. It consists of *a body and a process*. The *body* consists of a large portion externally placed, which presents *externally* a roughened surface to which ligaments are attached. *Internally* a very irregular aspect, having at its postero-inferior part a somewhat square projection which superiorly is roughed for attachment of the capsular ligament of the hock, posteriorly has a groove coated with fibro cartilage, over which flexor pedis perforans plays, internally and inferiorly roughened surfaces for ligamentous attachment. Anteriorly a slightly concave oblong articulatory facet extending from above downwards obliquely inwards for articulation with astragalus. The main portion of the body just above this process on the inner surface presents a small elongated facet looking upwards, inwards, and forwards, which is continuous round an acute articular margin with another facet looking downwards and slightly forwards. The rest of the inner surface running to a point inferiorly is roughened for attachment of interosseous ligaments. Along the anterior margin may be seen a small facet looking inwards, continuous with a large one upon the margin looking forwards, the rest of this margin is rough. The *inferior margin* is mainly occupied by an irregular slightly concave articulatory surface for os cuboides, this is continued slightly on to the internal surface at its inferior point as a small facet for union with astragalus. From the *superior margin* the process runs upwards. The *posterior margin* is continuous with the posterior margin of the process and both are rounded and roughened for attachment of the calcaneo-cuboid ligament. The *superior extremity* of the process is tuberous, slightly inclined forwards. Anteriorly it presents a smooth surface for a bursa, over which gastrocnemius externus plays, becoming inserted just behind, while still farther back this surface is smooth for the bursa situated on the point of the hock, where the expanded tendon of gastrocnemius internus forms a cap for it, becoming attached by a fibrous band running to the process on either side. The *anterior margin* is thinner than the posterior but rounded. The *external surface* is subcutaneous. The *internal surface* coated with fibro-cartilage, continuous with that on the oblique posterior part of the inner prominence of the body, over which the perforans tendon plays, bound down by the posterior annular ligament.

OS ASTRAGALUS is a square bone at the inner part of the upper row of the hock bones. It articulates superiorly with the tibia, posteriorly with calcis, inferiorly with cuneiforme magnum and cuboides. It has five surfaces. The *antero-superior surface* presents two prominent rounded articular ridges, running obliquely downwards and outwards. The *inner ridge* is the longest, and its inner side the straightest, and completely coated with cartilage. It extends beyond the outer, both anteriorly and posteriorly. The *outer ridge* curls slightly outwards at its inferior extremity, and is separated from the inner by a deep synovial fossa, which is shorter than either of the grooves, and generally presents a break in the articular cartilage centrally. Between it, the grooves, and the inferior surface, is a triangular depression, roughened for attachment of extensor pedis accessorius. The *posterior surface* of the bone is extremely irregular. Internally it presents superiorly a roughened prominence, and below this the posterior surface of the **tubercle**. More outwardly, separated from the ridge by a deep perforated fossa, is an ovoid articulatory surface for os calcis, extending directly downwards; and more outwards, inclined to the outer surface and extending obliquely forwards, is an articulatory facet consisting of two parts, one of which is vertical, the other horizontal. A little lower down, on the outer side, is another small facet, elongated from above downwards. We see a third at the infero-external angle of the bone continuous with that on the inferior surface of the bone. The interarticular space is rough and perforated. The *outer surface* is small, rough, and concave, giving attachment to one of the lateral ligaments. The *inner surface* is roughened and flat superiorly, inferiorly forms a prominent tubercle; to the whole of it ligamentous substance is attached. The anterior margin of both the inner and outer surfaces is occupied by the outer margin of the supero-anterior articular ridges. The *inferior surface* is mainly articulatory, consisting of a slightly convex surface, elongated from side to side, externally continuous with a small triangular facet for os cuboides, which is continuous with one on the posterior part for os calcis. Anteriorly it almost reaches the internal ridge, and at the under surface of the tubercle is reflected backwards, extending to about the centre of the posterior part of the bone, where it terminates in becoming continuous with the inner articulatory facet for os calcis of the posterior part. The anterior and reflected portions are separated by a gap in the articular cartilage, which presents numerous foramina. This portion internally is more concave than the rest of the surface, being separated from its outer extremity by a prominent point. This large articulatory surface comes in apposition with cuneiforme magnum.

OS CUNEIFORME MAGNUM is a flat bone with two surfaces and four margins. The *superior surface* is concave in most parts, and is prominent at the outer part of its anterior margin. It is the exact counterpart of the inferior extremity of astragalus (with exception of its small surface for os cuboides), and therefore requires no detailed description. Its posterior part is most prominent, especially inwardly; *inferiorly* it articulates with ossa cuneiforme parvum and medium. For medium it presents a large articulatory surface, extending along the anterior border, having a convex anterior, concave posterior margin almost flat, with its inner and larger half inclined slightly inwards. At the postero-external angle is another facet for medium, which is rounded, slightly concave, continuous with a small facet on the outer margin at the posterior part for os cuboides. For parvum there are two small round facets, continuous with each other at the antero-internal angle, the outer being concave, almost continuous with the anterior surface for medium, the inner convex looking downwards and backwards. The rest of this surface is extremely rough for attachment of interosseous ligaments; in some places it is perforated by foramina. The *anterior* and *internal margins* of the bone are roughened for attachment of ligaments, and the *posterior margin* presents two very irregular prominences separated by an irregular groove, which is continuous with the roughened groove upon the inferior surface. The posterior surface for articulation with cuneiforme parvum encroaches upon the inner prominence below. The outer prominence inclines slightly outwards, and on its postero-external part presents a concavo-convex surface, extending obliquely upwards and backwards for union with os cuboides. In connection with the external extremity of the anterior articulatory part of the upper surface is a facet elongated from before backwards, also coming into connection with os cuboides. Behind this is a groove extending upwards and backwards to the break in the superior surface. The rest of the surface is rough for attachment of interosseous ligaments.

CUNEIFORME MEDIUM is a flat bone, situated at the lower, inclined to the inner, part of the hock. It may be distinguished from cuneiforme magnum by a projection from the centre of its posterior margin, which is inclined slightly inwards. The *upper surface* is for the most part articulatory, presenting two distinct facets for union with magnum. The first is oblong with a notch in its posterior margin. It presents an anterior and a posterior rounded margin; the surface slants from the former to the latter; it is, therefore, concave. The posterior notch is separated by a roughened perforated space from an ovoid facet, extending obliquely inwards on the superior surface of the process. The *under surface* much resembles this, but the articulatory facets are convex and less regular, and the posterior facet is elongated from side to side, inclining outwards and backwards, also the intervening roughened space is larger. It articulates with os metatarsi magnum, and posteriorly with the inner splint bone. The *anterior margin* is convex, most prominent along its central line, and on the inner side, near both the upper and the lower surfaces are rows of foramina. The *inner margin* is a deep groove, the sides of which meet at an oblique angle. It is everywhere roughened, except at its antero-inferior angle, where is a small articulatory surface for union with cuneiforme parvum. The *external margin* presents a shallower groove than the internal, and a larger articulatory surface at the antero-inferior part for cuboides. The *posterior angle* at which these two surfaces meet looks obliquely downwards and backwards, being prolonged superiorly into a very thin process, with the superior part of which cuneiforme magnum articulates, and os cuboides with the external lateral part.

CUNEIFORME PARVUM is situated at the postero-internal and inferior part of the hock. It has two surfaces and a process. The *superior surface* is mainly occupied by a smooth concave articulatory surface looking inwards, bounded anteriorly by a roughened ridge, for articulation with magnum. The *inferior surface* presents three articulatory facets. The posterior facet is ovoid and distinct, and articulates with the inner small metatarsal bone, as also does the central facet, which is continuous with the anterior facet for metatarsi magnum, and that in its turn with a small surface for medium on the antero-internal part of the bone. The **process** is continued from the posterior margin, is flat from side, tuberous, has rounded margins, and curls inwards. It is roughened externally for attachment of ligamentous structure (especially a band running to os cuboides) as are also the interarticular spaces. The roughened depression on the inferior surface is sometimes so deep as to divide the bone into two distinct parts.

OS CUBOIDES occupies the outer part of the lower row of hock bones, and is equal in thickness to the sum of the two cuneiforme bones (magnum and medium). This thickness is characteristic, and serves to distinguish this from all other bones except os scaphoides of the knee, from which it may be distinguished by the greater number and the smaller size of its articulatory surfaces. It has six surfaces. The *anterior* and *posterior* are roughened for ligamentous attachment, more especially the anterior for the middle anterior annular ligament for which there is a slight depression. The *external surface* presents a groove running from below upwards, separating the roughened anterior and posterior parts. This leads upwards to a roughened line which tends to separate the upper surface into two articulatory parts and after reaching the inner surface curves in a direction backwards. Thus at the postero-external angle it cuts off a convex, ovoid facet sloping obliquely downwards, outwards, and forwards for union with os calcis. The anterior articulatory facet is divided into two parts, the outer largest, quadrilateral at the antero-external angle of the surface, slanting slightly outwards and forwards, continuous along its internal margin with the inner part which is pyriform (broadest posteriorly) for union with astragalus. It does not reach to the anterior margin of the bone, but along its inner edge blends with a pyriform facet, with its base anteriorly placed at the antero-superior angle of the internal surface for cuneiforme magnum. The *internal surface* at its postero-central part has a prominence, smooth and articulatory both above and below. The superior facet looks upwards and inwards meeting os cuneiforme magnum. The inferior, downwards and inwards for cuneiforme medium. At the antero-inferior angle is a facet elongated from before backwards continuous with one for metatarsi magnum on the inferior surface, it articulates with medium. The space between these is roughened for ligamentous attachment, but around the posterior and superior part of the last mentioned facet is a deep smooth groove extending to below the articulatory prominence posteriorly, through which the communicating branches between the anterior and posterior tibial veins run. The *inferior surface* presents two facets, one at the antero-external angle, the other at the postero-external angle for union with os metatarsi parvum externum. The former is largest and continuous inferiorly with a surface sloping upwards and inwards for metatarsi magnum, which joins one on the inner surface for cuneiforme medium. The *superior surface* is occupied by a smooth articulatory surface elongated from before backwards with the greater portion of which os calcis comes in contact.

The remainder of the bones of the hind limb resemble

the corresponding bones of the fore, though each presents certain points of difference which will be noticed.

OS METATARSI MAGNUM *differs from the large metacarpal bone.*—In being about one and one-third times as long, in being rounder, in having the articular surfaces for the splint bones more posteriorly placed. In a tendency to the production of a sharp ridge running from the upper surface of the bone downwards over the central line of of the antero-superior part. In the arrangements of its upper surface, which presents two distinct articulatory surfaces, one anteriorly crescent-shaped, one posteriorly placed rounded. These are separated by a roughened groove. Supero-externally is a groove running obliquely downwards and backwards for the metatarsal artery.

OS METATARSI PARVUM EXTERNUM is much the largest of the four splint bones. Its upper extremity presents at the antero-external part a facet for os cuboides, but the rest of this extremity which is rough and tuberous as well as its size renders it readily distinguishable.

OS METATARSI PARVUM INTERNUM is distinguishable by its upper extremity, which has two articulatory facets (continuous with that for union with os metatarsi magnum) for os cuneiforme parvum, and one on its outer side for cuneiforme medium. These facets with the rest of the surface which is rough, render the extreme upper part somewhat pointed.

OS SUFFRAGINIS is larger at its superior extremity, smaller at its inferior than that of the fore limb.

OSSA SESAMOIDEA, OS CORONÆ present no marked difference; **OS NAVICULARE** is smaller.

OS PEDIS may be distinguished:—By the slight obliquity of its antero-lateral surface; by the greater concavity, less circular, more regularly semilunar shape of its plantar surface, which is widest at the toe and gradually decreases in size towards the heels, while that of the fore limb present an indentation into the central part of its posterior crescentic margin.

PART III—ARTHROLOGY.

ARTICULATIONS are the unions of the component parts of the basement structures, the means by which such parts are enabled to play on one another in adaptation to the various movements of the individual. These are of three kinds, *moveable, partially moveable, and fixed*. The **fixed or synarthrotic joints** are those which in the young animal are moveable, but become obliterated later on in life; they mainly serve for connection; motion between the parts forming them is but slight, it is impossible after full development of the bones. They are of *four kinds*, the most marked examples of which may be found in the cranium. They consist of the coaptation of edges or surfaces of bones with the intervention only of periosteum. The simplest forms afford the most firm centres of connection.

Sutura is the apposition of two edges or surfaces, more or less complete, by the adaptation of prominences on one piece to depressions on the other. This adaptation may be by the *juxtaposition* of irregular surfaces, *sutura squamosa*, which may be more or less rough; examples of this may be seen in the union of the malar with the frontal bone; or it may be by the *dovetailing* of two jagged edges, *sutura serrata vel dentata*. The latter name is applied when the jagged margins resemble the line formed by the edges of a dog's molars.

Harmonia is the apposition of two edges of bone, which may be roughened, but are not jagged; they therefore simply are in contact, and connection between the bones is simply by continuity of periosteum, as in the union between two ossa nasi.

Gomphosis is the insertion of a peglike process of one bone into a corresponding cavity of another. It is rare; the connection of the teeth with the jaws and of the rib with its cartilage are examples.

Schindylesis is the union of three bones, by the attachment of the edge of one bone into a groove formed by part of the edges of the others, the rest of the union between the other two being by suture. Such is the attachment of the vomer to the palatine processes.

Partially moveable or amphiarthrotic joints are those which, while admitting a small amount of motion, also afford firm union. This is brought about by insertion between two roughened surfaces or edges of bone of a piece of fibro-cartilage, the perichondrium of which is continuous with the periosteum of the bones connected. Sooner or later, in different individuals, the cartilage becomes ossified and the union complete. Such is the connection of the ossa innominata by means of the symphysis, and of the two halves of the inferior maxilla, which "grow together," forming the symphysis inferioris maxillæ. In the case of the articulations between the bodies of the vertebræ we see permanent amphiarthrotic joints. Some of these joints are strengthened by connecting ligaments.

Diarthrotic, synovial, or freely moveable joints are those in which the component parts glide more or less upon each other. They present cavities lined by synovial membranes, the surfaces of the bones in apposition are coated with articular cartilage, and are maintained in apposition by ligaments. These ligaments may be *interosseous*, situated between the bones, running from a roughened spot on the surface of one bone to a corresponding part of its fellow; examples may be seen between many of the bones of the hock and knee; *connecting*, running from the outer surface of one bone to that of the other. *Capsular*, broad, expanded bands of white fibrous tissue serving to strengthen and protect the synovial membranes of a diarthrodial joint. *Annular*, extending round a complex joint attached to all the prominent bony points it presents, keeping in position its surrounding structures; generally consists of a broad ligamentous expansion, sometimes simply of a small white fibrous band; examples of both these forms may be seen in the hock. The motion between synovial surfaces may be simply an almost imperceptible gliding; where this is the case the joint is **arthrodial**. Others admit of more or less extensive motion in two ways—either abduction and adduction (motion from or towards the longitudinal vertical plane of the body), or elevation and depression (motion to or from the ground surface). These are **ginglymoid joints**, and in them motion may be limited by bony prominences sooner or later in either direction; the elbow-joint is an apposite example. **Axonoides or pivot-like motion** is motion on the longitudinal axis of the joint, as ginglymus is on the transverse axis; the union between atlas and axis is of this nature. **Rotatorial or ball-and-socket joints** allow flexion, extension, abduction, and adduction, axonoidal motion, and circumduction (whereby the above movements are manifested in combination); examples are the hip and shoulder-joints, in which a more or less globular head fits into a corresponding cavity. Joints are by no means constant; thus sometimes a small synovial sac may be found in the centre of a symphysis; some bones are united together by two kinds of joints; thus most of the vertebræ are united to their fellows both by amphiarthrosis and by diarthrosis.

The **ARTICULATIONS OF THE HEAD** are synarthrodial with few exceptions. They present every variety of this class, while the petrous temporal bone and os hyoides are united by amphiarthrosis, and the upper to the lower jaw by diarthrosis. The latter only presents unusual characters.

Temporo-maxillary articulation is the apposition and connection of the glenoid cavity of the squamous temporal bone with the condyle of the inferior maxilla. In birds there is an intervening bone, **os quadratum**, but in mammalia this is only represented by *an inter-articular disc of fibro-cartilage*, which serves to separate the cavity of the joint into two distinct parts, each lined by a synovial membrane, so that both surfaces of this cartilage are lubricated by synovia. It is a flattened piece of cartilage, elongated from side to side, and all around its outer margin is attached the *capsular ligament of the joint*, which extends from the outer circumferent margin of the glenoid cavity to that of the condyle. It is strengthened by a *fibrous band running* from the outer part of the condyle to the outer part of the cavity, and by *another* running from the posterior part of the condyle to the mastoid process of the squamous temporal bone, and continued from this up to the base of the external auditory process, covering a small synovial cavity. This joint internally is covered by pterygoideus, externally by masseter externus.

In examining the synarthrodial articulations of the skull we notice several sutures, one of which divides the bone into two parts longitudinally, only noticeable superiorly and not extending to the posterior extremity; this is the **saggittal suture** between the nasal, frontal, and parietal bones of the two sides of the head, forming the "plume" of the arrow posteriorly between the parietal and triquatral bones. Other sutures serve to mark the division of the skull into four vertebral segments; they are the

Lambdoidal, between os occipitis and os parietale, so called from its likeness to the Greek lambda (λ), which is caused by protrusion of the mastoid ridge of the petrous temporal between these two bones.

Coronal, between the frontal and parietal bones.

Great transverse, between the frontal and nasal, malar, lachrymal bones, extending into the orbit. All these extend downwards on either side towards the base of the skull.

Atloido-occipital articulation

is the union of the condyles of the occiput with the anterior articulatory surfaces of the atlas. Each of these condyles has a proper synovial membrane, and the two lie together on their inner and inferior parts. Their outer surfaces are *strengthened by accessory fibres*, which are continued superiorly, extending from the anterior margins of the laminæ of the Atlas to the superior part of foramen magnum of the occiput, forming the capsular ligament. *Two flat, white, fibrous bands* extend from the styloid processes of the occiput to the anterior margins of the laminæ. This joint is ginglymoid, admitting motion upwards and downwards.

Atlo-axoid articulation

is the union of the posterior part of the body of the atlas and the posterior part of its superior surface with the under surface of the odontoid process, and its antero-lateral continuations on the dentata. These have one synovial membrane attached to the outer circumferent margin of both surfaces. Inferiorly along the central line this is strengthened by a *small band of yellow elastic tissue* (a continuation of the inferior vertebral ligament). Laterally it has *accessory fibres, which are continued upwards*, connecting the anterior laminal margin of the axis with the posterior laminal margin of the atlas. The superior spinous process of the dentata is connected to the upper surface of the laminæ of the atlas by *yellow elastic bands lying together along the middle line* expanded in becoming blended with the interlaminal ligament outwardly. The superior surface of the synovial membrane is in contact with the **odontoid ligaments**, three in number:—The *broad ligament* runs from the transverse ridge on the upper surface of the odontoid process, the **two small ligaments** are beneath this just visible, one on either side, running from the depressions on the upper surface of the odontoid process. The three run to the roughened surface at the antero-superior part of the body of the atlas. This is a specimen of an *axonoides* articulation; it allows motion of the head from side to side, whereby the muzzle may be made to describe a circle.

Vertebral articulations.

The rest of the vertebral articulations have many characters in common. The anterior convex surface of the body is united to the posterior concavity of the body in front of it, through the medium of an intervertebral disc of cartilage, which is soft and cartilaginous centrally, hard and fibrous externally. Inferiorly its fibres in some cases take a marked direction from before backwards along the central line, forming the **inferior vertebral ligament**, most marked in the lumbar region, where it affords attachment to the crura of the diaphragm. The **superior vertebral ligament** serves to strengthen this union superiorly, for it runs along the floor of the spinal canal, its whole length consisting of a series of bands attached to the roughened upper surfaces of the vertebral centra, widest in becoming attached to the intervertebral discs, growing narrow opposite the centres of the vertebral bodies, in forming a small band passing over the transverse groove for a venous sinus.

The anterior oblique processes of the vertebræ are united to the posterior oblique processes of the vertebræ in front, each by means of a synovial membrane with a strengthening **capsular ligament**. This articulation is *arthrodial*, and is separated inferiorly by the intervertebral gap from the amphiarthrodial union of the bodies, supero-internally from its fellow by the *interlaminal ligaments* which, running from the posterior margin of the laminæ of the anterior vertebra, gain the anterior margin of the laminæ of the posterior vertebra, extending inwards as far as the *interspinous ligament*, which runs from the posterior margin of the superior spinous process of one vertebra to the anterior margin of the spine of the next bone. This is not found in the cervical and coccygeal regions, and with the interlaminal becomes converted into bone in the sacral.

The *superspinous ligament* is a dense band of white fibrous tissue extending along the upper extremities of the superior spinous processes from the sixth dorsal to the coccygeal vertebra. Posteriorly it becomes gradually lost. In the sacral region it affords attachment to the important muscles of the quarter, triceps abductor femoris, biceps rotator tibialis, and ischio tibialis; and in extending downwards on either side of the spinous processes forming that flattened band which becomes attached to the oblique processes termed the **sacral ligament**, gives attachment to longissimus dorsi and to some of the muscles of the tail. It is very intimately blended with the superior ilio-sacral ligament and with the tendinous layer of longissimus dorsi against the posterior spine of the ilium. In front of this it gives attachment laterally to longissimus and latissimus dorsi and to other muscles through the medium of the lumbar faschia. In the cervical region it is represented by

Ligamentum nuchæ. This is a remarkable body, wholly composed of yellow elastic tissue, situated in the middle plane of the neck, consisting of three portions, superspinous, cordiform, and membraniform. The *superspinous portion* consists of double bands extending between the superior spinous processes of the cervical and four anterior dorsal vertebræ; they are expanded inferiorly, blending with the interlaminal ligaments. Superiorly they blend with the *membraniform portion* which consists of two thin layers of closely and somewhat irregularly interlaced elastic fibres. These layers rest against each other by their inner surfaces, inferiorly are formed by a number of bands running upwards and backwards from the cervical and anterior dorsal spinous processes. Thus the muscles which occupy the spaces between the superior spinous processes in this region lie in contact with their fellows along the central plane, but those more superiorly placed are separated by the two layers of the membraniform portion of this ligament. The muscles mainly thus separated are the complexi majores. The superior margin of this portion blends with the **cordiform portion**, which is a rounded band of yellow elastic tissue, extending from the superior extremity of the spine of the fifth dorsal vertebra (where it is directly continuous with the superspinous ligament) to the postero-inferior part of the crest of the occiput and the line between its scabrous pits. To its upper surface, which is flattened, is attached the fatty mane, and on either side it affords more or less direct attachment to panniculus carnosus, levator humeri, splenius, trapezius, rhomboideus and complexus major. This elastic ligament serves to support the head without muscular exertion.

COSTAL ARTICULATIONS.

The upper extremity of the rib presents three synovial surfaces for union with the spine. The **tubercle** *articulates with the transverse process of the vertebra behind*, by one facet which has a *synovial membrane* (strengthened to form a *capsular ligament*) and *two costo-transverse lateral strengthening bands*, one externally, the other internally placed. The **head of the rib** presents two facets separated by a roughened groove. The facets unite by diarthrosis with corresponding facets, one on the postero-superior and lateral part of the body of the vertebra in front, the other with the antero-superior and lateral part of the vertebra behind. Each of these facets has a synovial membrane strengthened to form a *capsular ligament*, and from the groove between *ligamentum teres* runs to the intervertebral disc. From the inner surface of the neck of the rib the *stellate ligament* runs, and divides into three parts, one of which runs to the body of the vertebra in front, another to the body of the vertebra behind, the third to the intervertebral disc of cartilage.

The gomphotic union of the inferior extremity of the rib with the sterno-costal cartilage is maintained by continuity of the costal periosteum with the investing perichondrium of the cartilage. The *inferior extremity of the cartilage of elongation* has a diarthrodial union with the sternal cartilage or bones in the case of all the true ribs; it presents a synovial membrane and *two lateral ligaments*, the superior of which blends with a *fibrous band which runs from one extremity of the sternum to the other*, and posteriorly is continuous with the ligament by which the inferior sharpened extremities of the false ribs are connected together.

ARTICULATIONS OF THE FORE EXTREMITY.

Shoulder or scapulo-humeral articulation

is the union of the glenoid cavity of the scapula with the head of the humerus; presents only a *capsular ligament*, directly connecting the outer circumferent margins of these two articular surfaces. It is lined by synovial membrane. Anteriorly its external surface is separated by fat from the tendon of flexor brachii, and the bursa between it and the antero-superior part of the humerus. The muscles in contact with this surface are:—

Internally—subscapularis; posteriorly, caput magnum, scapulo-humeralis posticus; externally, scapulo-humeralis externus, postea-spinatus, and antea spinatus.

This is a good example of a *rotatorial* joint.

Elbow or brachio-humeral articulation

is the union of the condyles of the humerus with the articulatory superior surface of the radius and that of the hamular process of the ulna. This articulation has a break in the articular cartilage of both of the articular surfaces. It presents a *capsular ligament* and *two lateral*. The *outer lateral ligament* is the shortest but stoutest. It runs from the depression on the external surface of the outer condyle to a slight prominence at the superior part of the outer margin of the radius. The *inner lateral ligament* superiorly is attached to a depression on the inner surface of the inner condyle, runs downwards, is attached to a prominence at the superior part of the inner margin of the radius, passes over the tendon of humeralis externus, which it binds down, and becomes attached below it.

It is a *ginglymoid* articulation.

Radio-ulnar articulation

has been necessarily somewhat minutely described in our examination of its component bones.

The knee-joint, carpus, wrist of human subject,
consists of four distinct joints, all synovial and having
distinct capsular ligaments. The **radio-carpal joint** is
between the inferior extremity of the radius and the scaphoid, lunar, and cuneiform bones and trapezium. These four
bones are united together by *interosseous* and *connecting
ligaments*, which it is unnecessary to specify. The **carpal
joint** is between the row thus formed, and the lower row
formed by trapezoides, magnum, and unciforme united
together in a corresponding manner, which by their under
surfaces come into connection with the upper surface
of the metacarpus, thus forming the **carpo-metacarpal
joint.** The last joint is *arthrodial*; the two former *ginglymoid*;
each has a synovial membrane, which in every case dips
down between the small facets presented by their small
bones for union with one another, which have been already
minutely described. The *fourth capsular ligament* is found
surrounding the synovial joint which trapezium makes
with the cuneiform bone. These different joints are connected by *lateral ligaments and the annular ligaments*. The
outer lateral ligament may be divided into four, the *inner*
into three parts. The **annular ligament** is attached
anteriorly to the roughened anterior surfaces of all the
small bones of the knee, also to the roughened supero-
anterior part of the metacarpus. It *anteriorly presents
three channels* running from above downwards with special
synovial thecæ; the *inner one* extends obliquely downwards
to the inner small metacarpal bone and gives passage to
extensor metacarpi obliquus. The *outer one* runs straight
downwards for passage of extensor pedis. The *central* is
broadest and terminates opposite the supero-anterior part
of metacarpi magnum, where the extensor metacarpi
magnus, to which it gives passage, is attached. A continuation of the annular ligament surrounds extensor suffraginis in the major portion of its course from the elbow
to the knee, and after it has passed through a sheath
formed by the outer part of this ligament, from which it
receives a *reinforcing band, it is termed ligamentum extensorum*. On the surface of trapezium it forms a groove
with a theca for the lower tendon of flexor metacarpi
externus. The arrangement of the posterior part of the
annular ligament with its synovial membranes, and its

superior suspensory and *subcarpal limiting bands* has been minutely described in connection with the tendons of the flexores pedis.

Fetlock joint

is the union of the inferior extremity of the cannon bone superiorly with the ossa sesamoidea and os suffraginis inferiorly. We, therefore, must examine how the four bones are united together. The sesamoid bones are covered posteriorly by fibro-cartilage, for they form a groove over which the flexor tendons play. This cartilage serves to connect the two sesamoids together in passing between them, forming the *intersesamoid ligament*, and superiorly it is prolonged above the bones, continuous with two small bands of a decidedly elastic nature running to the postero-inferior part of the large metatarsal bone. The sesamoids are connected to os suffraginis laterally by broad bands of white fibrous tissue, *lateral sesamoid ligaments*, inferiorly by several ligaments, the **inferior sesamoideal**.

The *superficial inferior sesamoid ligament* consists of two lateral and a central portion, all of which extends from the supero-posterior part of os suffraginis to the inferior part of the mass formed by the sesamoids and their fibro-cartilage. The *middle ligament* has a widespread triangular attachment to the posterior surface of os suffraginis and extends to the inferior part of the sesamoid mass. The *deep ligament* rises from the small triangular space at the supero-posterior part of os suffraginis, sends fibres which cross each other before reaching the sesamoid mass. The combined sesamoid and suffraginis bones are attached to the large metacarpal bone by the

Superior sesamoideal or suspensory ligament which extends from the supero-posterior part of os metacarpi magnum and as a firm fibrous cord, with more or less numerous muscular fibres interspersed, runs down the back of the limb in the groove formed by the posterior surface of os metacarpi magnum and the small metacarpal bones on either side. Opposite the superior part of the inferior third it bifurcates, one division running to the outer part of each sesamoid, and from this being continued onwards over the lateral parts of os suffraginis to blend with extensor pedis on the anterior part. This ligament corresponds to a number of muscles found in this situation in animals which have five digits; sometimes these muscles are found so marked in the horse as to receive the name **lumbrici**. The **lateral ligaments of the fetlock** extend from the infero-lateral part of metacarpi magnum almost as high up as the inferior bulbous prominence of the splint bone, downwards to the supero-lateral part of os suffraginis. They are connected together by the **capsular ligament**, a thick layer of white fibrous tissue attached to the large metacarpal and suffraginis bones close to their articulatory surfaces ; internally it is lined by the synovial membrane of the joint, externally in contact with extensor pedis tendon, and extensor suffraginis. The annular ligament passes from the lateral ligaments behind the flexor tendons, and inferiorly has two bands which blend with perforatus tendon at its bifurcation.

PASTERN JOINT.

The inferior articular surface of os suffraginis is maintained in apposition with the superior surface of os coronæ, anteriorly by the tendon of extensor pedis, posteriorly by the long inferior sesamoideal ligament. On each side is a stout **lateral ligament**, superiorly extending almost half way round the bone from underneath extensor pedis, anteriorly to the external part of the posterior surface. Its fibres run downwards and backwards, and after covering the inferior attachment of flexor pedis perforatus become attached to the supero-lateral parts of os coronæ, some of them being continued on to form one of the divisions of the stellate navicular ligament, others joining with similar fibres from the other side to form a thin fibrous layer, the superior surface of which intimately blends with the perforans tendon ; the inferior gives attachment to the fibrous frog.

The **coffin or pedal joint** will hereafter be described. (*See Foot.*) It has only two ligaments, *lateral*.

LIGAMENTS OF THE PELVIS.

The superspinous ligament in the sacral region is very thick and strong, and anteriorly expands, becoming attached to the postero-superior spinous process of the ilium. It is here continuous with the tendinous structure of longissimus dorsi, and is termed the *superior ilio-sacral ligament*. Becoming attached along the posterior margin of the ilium, extending from the postero-superior spinous process about three inches downwards, is the *inferior iliosacral ligament*, the fibres of which run obliquely downwards and backwards, to become attached inferiorly to the transverse ridge of the sacrum, and to become continuous onwards, blending with the substance of the sacrosciatic ligament, after forming with that ligament below the transverse ridge a canal extending from before, backwards, through which an important branch of the gluteal artery, with its accompanying vein and nerve, pass. The space on the lateral part of the sacrum bounded outwardly by the inferior ilio-sacral ligament and by the superior part of the venter ilii is occupied by the posterior part of longissimus dorsi, and when this is removed, a continuation of the superior spinous ligament, under the form of a thick layer of fibres running downwards and backwards over the lateral surface of the sacral spines, is exposed. It is termed the *sacral ligament*.

The *sacro-iliac ligament* consists of a number of stout ligamentous fibres, diverging in all directions from the roughened surface around the auricular facet of the first sacral transverse process. They run to the roughened surface of venter ilii, and externally blend with the inferior ilio-sacral ligament, internally being continuous with a stout ligamentous band running towards the articulatory margin on the anterior part of the last lumbar transverse process, to the summit of which also a few scattered fibres run.

The **sacro-sciatic ligament** forms the postero-lateral boundary of the pelvis. Superiorly it is attached firmly along the whole length of the transverse ridge of the sacrum, inferiorly it becomes attached to the ischiatic spine, and to the tuberosity of the ischium, from the line between these two prominences becoming reflected inwards as *a fibrous layer between the pelvic viscera, and the internal obturator and pyriformis muscles* becoming attached to the superior edge of the posterior margin of the ischium and along the upper part of the inner edge of the os innominatum. Thus between the ligament and the bone are two spaces; the **superior-ischiatic notch** is between the anterior margin of the ligament and the posterior margin of the ilium, and gives passage to the gluteal vessels and nerve and the sciatic nerve. The **inferior-ischiatic notch** between the neck of the ischium and this ligament, where it is reflected inwards, gives passage to the obturator internus and pyriformis tendons; the external surface of this ligament gives attachment to some of the large muscles of the quarter and over it, almost embedded in it, runs the great sciatic nerve. Its inner surface affords attachment to retractor ani and compressor coccygis. The **obturator ligament** is an extremely thin layer of white fibrous tissue, with difficulty demonstrable, situated between the internal and external obturator muscles, attached to the margins of the foramen, deficient at the antero-external part, where the obturator vessels and nerve gain passage.

ARTICULATIONS OF THE HIND EXTREMITY.

Hip-joint

is formed by the head of the femur and the acetabulum, maintained in apposition by ligaments. The acetabulum is increased in depth by the *circumferential or cotyloid ligament*, a ring of fibro-cartilaginous substance attached around its margin, it passes over the deficiency on the inner side of the cavity, and here is termed the *transverse ligament;* thus it leaves a space through, which a dense ligamentous band of a tendinous nature runs to become attached to the break in the articular cartilage on the head of the femur. This is the *pubio-femoral ligament*, a continuation of the common tendon of the abdominal muscles; it runs from the anterior part of symphysis pubis outwards through a groove on the under surface of os pubis; some of the muscular fibres of rectus abdominis are directly inserted into it; at its attachment to the head of the femur it blends with *ligamentum teres*, which passes from here to the internal roughened part of the acetabulum. From the free margin of the cotyloid ligament the *capsular ligament* extends downwards, to become attached around the outer margin of the articulatory surface of the head of the femur; it is lined by synovial membrane, which also invests the pubio-femoral and teres ligaments.

Stifle joint

corresponds to the knee of the human subject; the patella, tibia, fibula, and femur assist to form it. Between the condyles of the femur and the upper articulatory surface of the tibia on either side is a **semilunar disc of fibro-cartilage.** These are convex and thick externally and decrease in size to form a thin edge at the internal margin around which the synovial membrane is reflected from the upper to the under surface; to the outer margin the capsular ligament is firmly attached. The anterior extremity of each cartilage is attached by a ligament to the anterior part of the spine on the upper extremity of the tibia. The posterior extremity of each by a ligament to the upper extremity of the tibia behind the spine, but the *outer cartilage* in addition has a ligament extending to the posterior part of the inter-condyloid fossa of the femur. The femur is also connected to the tibia directly by the crucial and the lateral ligaments. The **crucial ligaments**, two in number, can only be exposed by removal of one of the condyles of the femur. *One* of them runs from the anterior part of the groove on the tibial spine, to become attached supero-posteriorly to the inner surface of the outer condyle in the intercondyloid fossa; its fibres cross at right angles, *the other crucial ligament*, which runs from the anterior part of the intercondyloid fossa downwards and backwards to the superior extremity of the tibia behind the spine.

The *outer lateral ligament* is superiorly attached to the depression on the outer surface of the outer condyle of the femur; it crosses the tendon of popliteus which plays through a sheath between this and the capsular ligament, and then passes on to the femur. This ligament becomes attached below to the upper extremity of the fibula; from it peroneus mainly arises.

The *inner lateral ligament* is longer and narrower than the outer, extending from the internal surface of the inner condyle downwards to the supero-internal part of the tibia.

The patella is connected to the femur by **two lateral ligaments.** The inner extremity of the patella is continued as a piece of fibro-cartilage which plays over the inner condyle, and from this the *inner ligament* runs to the inner surface of the internal femoral ridge. It is smaller than the *outer*, which runs from the outer extremity of the patella to the corresponding part of the outer ridge.

The patella is connected to the tibia by **three straight ligaments.** The *inner* runs from the complementary cartilage on the inner side of the patella to the supero-internal angle of the ridge on the antero-superior part of the tibia. The *outer*, which is composed of two portions, extends from the outer extremity of the patella to the supero-external angle of the same ridge. The **capsular ligament**, after connecting together the two articulatory surfaces posteriorly and laterally, and internally becoming attached to the semilunar cartilages, externally to the lateral ligaments (partially separated from the outer by the tendon of popliteus), blends with the inner and the outer straight ligaments of the patella, connecting them together, and covering the *central straight ligament*, which commences about the centre of the anterior surface of the patella, plays over a bursa on that bone, is separated from the main joint by a quantity of fat, plays over another bursa at the central part of the superior extremity of the tibial ridge, and becomes attached below it.

The *synovial membranes of this joint are two or three in number*, for in some cases we see the cavity of the joint between the femur and the patella communicating with that of the inner condyle of the femur. The cavity of the outer condyle always remains distinct. These cavities frequently communicate with the important bursæ in the neighbourhood. The condyles come *directly* into contact with the lateral parts of the spine of the tibia, the upper surfaces of the semilunar cartilages are concave to accommodate the convex surfaces of the condyles, their under surfaces are plane.

The mode of union of the fibula to the tibia has been already described.

HOCK JOINT.

The small bones of the hock, like those of the knee, are united together by *interosseous and connecting ligaments*, thus producing two rows, the superior formed by astragalus and calcis, the inferior by the cuboid and three cuneiform bones. These, together with the metatarsus, form the *tarsal* and *tarso-metatarsal gliding joints*, lubricated with synovia secreted by membranes, which extend to the facets situated between the several small bones; generally a distinct synovial sac may be distinguished between the component bony layers of the lower row.

The **true hock-joint** is the ginglymoid articulation formed by the apposition of the inferior grooves of the tibia with the superior ridges of astragalus. This joint presents a *capsular ligament*, attached around the edge of the articulatory surfaces, which at the postero-internal part is thickened and fibro-cartilaginous, its external surface being here concave and lubricated with synovia, forming part of that theca in which the perforans tendon glides.

The *lateral ligaments of the hock* are two in number, and extend from the malleoli of the tibia to the supero-lateral parts of the large and respective small metatarsal bones; the *outer* is the largest. Both of them may be divided into several parts.

The **posterior annular ligament** is thickened in covering the bones on the postero-internal part of the joint in the groove, in which the perforans tendon plays; it likewise becomes reflected from the posterior border of os calcis to blend with the posterior margin of the inner lateral ligament, binding down the perforans tendon and continued on to retain the tendon of flexor pedis accessorius in its groove behind the internal malleolus.

The **anterior annular ligament** becomes marked as *three bands*. The *superior* extends across the antero-inferior part of the tibia, over the flat surface, binding down the tendons of flexor metatarsi and extensor pedis, inclined obliquely from within outwards and downwards, and continued on to retain the tendon of peroneus in its groove on the outer malleolus of the tibia. The *middle band*, extending from the cuboid, winds round extensor pedis tendon, and then blends with that division of the flexor metatarsi tendon which passes to the same bone. The *inferior band*, closely connected in the fresh subject with the middle, runs from the superior extremity of the outer small metatarsal bone over extensor pedis tendon to the central line of the supero-anterior part of the large metatarsal bone.

The **calcaneo-cuboid ligament** extends from the posterior border of os calcis along its whole length downwards to the superior part of the outer small metatarsal bone. Externally it blends with the external lateral ligament, internally with the posterior annular ligament. A large and important synovial membrane covers the inner surface of the capsular ligament.

None of the other ligaments of the hock require special notification.

The **other articulations of the hind limb** resemble those of the fore limb, to the description of which reference may be made when the dissection arrives at this stage.

PART IV.—SPECIAL ANATOMY.

The Head.

FROM the angle formed at the meeting of the incisions through the skin downwards from the ear, and backwards from the lower lip, the operator commences to remove the skin in the antero-superior direction as far as the root of the ear, the eyelids, and the central line of the face; this done he works forward towards the lips and the nostril, where the muscular fibres are found closely in contact with the skin, in some places being inserted into it. The roots of some of the long hairs which form the "feelers" will be also cut through in removing the skin from the lips. From the zygomatic ridge to the centre of the submaxillary space the **panniculus** in a thin layer lies immediately in contact with the skin, and through it may be seen the branches of the facial nerve, and the masseter externus muscle. Its fibres converge from here towards the angle of the mouth, where they blend with other muscles, especially orbicularis oris, forming the muscular band termed **retractor anguli oris**.

To the same point, and to the upper lip above it, run fibres forming a thin layer extending from the zygomatic ridge, and some from the muscles of the lower eyelid. This is **zygomaticus**, which muscle forms the central part of the superficial subcutaneous layer of the face.

The upper part of this layer is formed by **levator labii superioris alæque nasi**, which is attached superiorly to the nasal, frontal, and lachrymal bones at their point of junction, and runs downwards and forwards to be inserted into the point of the false nostril, the wing of the nostril, and the upper lip, blending with the other muscles of this part.

The fibres of this muscle are separated inferiorly into two

portions by the passage of the insertion portion of **retractor labii superioris**, which spread out in their course towards the ala of the nostril and the upper lip, while their origin from the superior maxillary bone in front of the termination of the zygomatic ridge is hidden by zygomaticus.

Extending round the free margin of the lips are a number of fibres of red muscular substance into which all the muscles running to these organs seem to pass; this is **orbicularis oris**, the muscle by which the lips are closed and the animal enabled to grasp small articles (as oat grains) on a flat surface. This muscle is nowhere intimately attached to the bone, but is composed of a series of short fibres encircling the oral opening.

The frontal and nasal bones are immediately in contact with the skin on the forehead and nose, but about opposite to the nasal peak the tendon of **nasalis longus labii superioris** emerges from beneath lev. labii sup. alæque nasi, and running over the cartilaginous prolongation of the nasal bones, where it is lubricated with synovia, it gains the anterior surface of the alæ of the nostril, which are here to be seen covered by **dilator naris anterior** (which runs from the external surface of one ala to that of the other); here the tendon blends with the corresponding tendon from the other side of the face, and from the point of union the fibres radiate to blend with orbicular oris, and become inserted into the skin of the upper lip.

The **false nostril** fills the angle formed between this tendon and the levator labii sup. alæque nasi. From the anterior edge of the nasal bones, extending as far back as where this bone meets the anterior maxillary bone, the **dilator naris superior** arises, and runs to be inserted into the false nostril along its whole superior margin.

Below retractor anguli oris the terminal portion of **retractor labii inferioris** with a small round tendon is found going to blend with orbicularis oris in the lower lip (this tendon is connected to the inferior maxillary bone below the corner incisor by a distinct band of muscular fibres in some subjects), and below this the bone is subcutaneous.

On removal of the superficial muscular layer of the face we expose the **masseter externus**, a huge square mass of muscle arising along the whole length of the under surface of the zygomatic process, and passing to the whole external surface of the branch of the inferior maxilla. At its origin it is tendinous, and through its substance run strong tendinous bands, which are inserted below into irregular ridges; these bands are a provision for the constant closure of the jaws when not required to be separated in mastication or otherwise, *for this muscle, with temporalis, pterygoideus, and masseter internus serve to close the mouth.* In many other situations we shall observe tendinous bands running throughout the whole length of muscles to support weight, without the active contraction of muscular fibre which would weary.

Along the posterior part of the superior margin of masseter externus we see the **temporal artery** with an accompanying vein, which vein is superiorly placed. This artery is one of the three terminal trunks of the external carotid, and after winding round the angle of the jaw posteriorly, proceeds for about two inches on the external surface of this muscle, and then dips down into its substance; the pulse is frequently taken at this artery. Along the anterior margin of the masseter externus run the **submaxillary artery** *and vein, with the duct of the parotid salivary gland.* The artery is almost invariably the most anteriorly situated, but the position of the vein and duct is not constant. At first in contact with the inferior maxillary bone they next pass over retractor labii inferioris, when the artery gives off a branch running forwards, the *inferior labial* and several branches, the *anterior masseteric*, which run from it posteriorly to the masseter muscle. They then run on to buccinator and caninus, the muscles which form the cheek; about the centre of this muscular mass the duct dips between the muscular fibres, and having proceeded a short distance beneath the buccal mucous membrane, opens into the mouth at a small eminence or papilla situated opposite the space between the second and third molar teeth (but the position of this opening is not constant). At the anterior extremity of the zygomatic spine the submaxillary artery divides into the *superior labial* and *facial.* This division occurs just after the vessel has passed over the originating tendon of retractor labii superioris; the *superior labial* branch runs forward, being mingled with the fibres of the superior maxillary division of the fifth cranial nerve, which here emerge from the infra-orbital foramen, and run to the upper lip. With these it pierces the lip, and so gains the incisive foramen, where it combines with the corresponding artery from the opposite side, and the trunk thus formed joins with the arch formed by the union of the two palatine arteries, which run along the sides of the roof of the mouth. The *facial artery* runs upwards, anastomosing with branches of the ocular arteries, which bend round the margin of the orbit. At its upper part it passes over the origin of nasalis longus labii superioris (which arises from about the point of junction of the nasal, malar, and lachrymal bones).

By removal of zygomaticus and retractor anguli oris we expose buccinator and caninus. These two muscles are intimately blended; **buccinator** anteriorly blends with orbicularis oris, its fibres run directly backwards to become inserted into the anterior margin of the ramus of the lower jaw, posteriorly to the alveolar cavities; it is also attached slightly to the alveolar ridges and to the tuberosity of the superior maxilla. To show this an incision must be made through masseter externus, whereby it will be found to cover superiorly a *varicose vein* situated immediately beneath the zygomatic spine, running from the submaxillary vein on the cheek through the foramen lacerum orbitale to the cavernous sinus; below this is one of the *buccal salivary glands* from two to three inches in length, similar in structure to the parotid gland, the secretion of which is poured into the mouth through numerous foramina opening upon a line of papillary eminences of the buccal mucous membrane, opposite to the upper molar teeth. Below this gland is seen buccinator intimately blended with retractor labii inferioris, inserted as above mentioned. **Caninus** is composed of two distinct layers of arched fibres, which arise from the superior and inferior maxillary bones respectively at the interdental spaces, run towards the central line of the anterior part of buccinator, where they blend together and with that muscle, sending some straight fibres to blend with it posteriorly. By raising the infero-anterior part of buccinator, just in front of masseter externus, we expose the **inferior buccal glands**, in every respect, except situation, resembling the superior; they lie in immediate contact with the buccal mucous membrane, with which the internal surface of buccinator is intimately attached throughout the rest of the cheek. Passing around the anterior margin of the coracoid process of the inferior maxillary bone is the *buccal branch of the inferior maxillary division of the fifth cranial nerve;* it sends some large fibres to the buccal glands, and runs onwards to supply buccinator and caninus as far forwards as the angle of the mouth. The *facial branch* of the same nerve winds round the posterior margin of the lower jaw just below the condyle, combines with the seventh nerve or portio dura, and proceeds over the cheek on the external surface of masseter externus. One of its portions runs with the temporal artery, the inferior branch passes to the lower lip, and there unites with that portion of the dental branch of the inferior maxillary nerve, which passes out of foramen menti (or the anterior maxillary foramen). The middle and largest branch proceeds over the cheek, and sends communicating fibres to the nerve which passes out of the infra-orbital foramen; these pass in a curved manner around the origin of retractor labii superioris, but the greater number of the fibres blend with the infra-orbital fibres, and run to the nostril and upper lip. Some small muscles on the muzzle remain yet to be noticed.

Dilator naris inferior arises from the superior process of the anterior maxillary bone, winds round its smooth superior margin, and becomes inserted into the cartilaginous anterior extremity of the inferior turbinated bone into the skin of the false nostril. *From the superior extremity of the false nostril a small muscle running to the superior maxillary bone just above the termination of the zygomatic ridge may frequently be found.*

Depressor labii superioris arises from the external surface of the body of the anterior maxilla around the alveolæ of the incisors, and extending back to the canines: it becomes inserted into the mucous membrane lining the upper lip, and may be exposed by removing this membrane and the labial glands. It will be found to blend centrally with its fellow. In the lower lip we find a corresponding muscle, **levator labii inferioris.** Internally to these muscles is a layer of salivary glandular substance from which run numerous ducts which pierce the mucous membrane of the lips, by inversion of which their openings may be rendered visible to the naked eye. The *structure of the lower lip* is completed by a mass of fatty matter throughout which many muscular fibres are interspersed. The **lips**, therefore, consist of skin, muscle, glandular structure, and mucous membrane, with veins, arteries, lymphatics, and especially *nerves*, the large size of which renders the "twitch" such an effectual "sedative?" It is orbicularis oris which forms the greater part of the lips, and which brings the two lips into close connection at their extremities, whereby the *commissures, or angles of the mouth,* are formed. The **lips** are the organs of prehension of the horse, and present an external surface from which long hairs, "*feelers,*" project, more especially from the inferior central prominent portion commonly called the chin. The internal surface of the lips is concave, moulded upon the anterior surfaces of the inferior and anterior maxillary bones (which is covered by the dense mucous membrane which forms the gum) and upon the anterior surfaces of the crowns of the incisors. The free margins of the lips are in contact with each other, or separated according as the mouth is shut or open.

The remainder of the skin may now be removed, when another orbicularis muscle will be exposed in the eyelids, which it tends to approximate and to maintain in apposition. This is **orbicularis palpebrarum**, which is attached by a small tendon to the lachrymal tubercle situated on the facial surface of the lachrymal bone. Its fibres extend into the upper and the lower lids, and meet at the *outer angle, commissure, or canthus of the eye.* It is externally in contact with the skin, which is thin and smooth in this neighbourhood. Internally it is fixed to the bones which form the rim of the orbit by loose areolar tissue, and in the substance of the upper eyelid is attached to the **tarsal cartilage or ligament**, a small semilunar portion of yellow elastic cartilage which is thickest in the centre, and grows narrower and thinner until it terminates in a point at either extremity in becoming continuous with the white fibrous tissue of the eyelid. It is sometimes found in the lower lid, but is then mostly fibrous. On the internal surface of this cartilage, and running in small parallel lines visible through the mucous membrane lining the eyelid, in a direction perpendicular to its free margin, are the **meibomian glands**, specially modified sebaceous glands, the product of which serves to prevent overflow of tears on to the external surface of the lids. To orbicularis palpebrarum run several muscles.

Levator palpebræ superioris externus runs from the

root of the orbital process of the frontal bone to the upper eyelid, where it blends with orbicularis, from the lower part of which muscle levator labii superioris alæque nasi and zygomaticus in part take their origin. The **eyelids**, therefore, are two semilunar folds which are adapted to cover the anterior surface of the eyeball. They are composed externally of skin, internally of mucous membrane, named the **conjunctiva**, which becomes reflected over the tunica albuginea and the cornea, thus covering all that portion of the eye which is naturally visible, and serving to retain the eyeball in its proper relation with the palpebral fissure, by its smooth surface to facilitate the movements of the eyelids, and to protect the eye. That portion of it which lines the eyelids and covers the tunica albuginea is vascular, but its continuation over the cornea is simply a basement membrane, with a single layer of tesselated epithelial cells, which is transparent and thus causes no obstruction to vision. Between these two layers of the eyelid are situated muscular structure, vessels, nerves, the meibomian glands and the tarsal ligament. The free borders of the eyelids are broad and only come into apposition by their outer edges when the lids meet; thus a triangular canal is formed by which the tears are directed towards the inner angle of the *palpebral fissure*, the space between the eyelids. Along the outer edge of this margin we see the **eyelashes, ciliæ**; long curved hairs, those of the upper lid, the strongest and longest, run in an upward direction, those of the lower lid downwards. This margin also presents in each eyelid a row of openings of the meibomian glands, and at the inner extremity a much larger opening; these latter are the **puncta lachrymalia**, through which the tears pass on their way from the front of the eye to the nasal chamber. The points at which the eyelids meet are termed the **canthi**; the *outer or temporal canthus* is in the form of an acute angle, and here the eyelids are in direct contact with the eyeball; *the inner or nasal canthus* presents a rounded diverticulum in which is situated a small rounded nodular body covered by a very thin dark skin which presents a few short and fine hairs. This is the **caruncula lachrymalis**, which is merely a mass of connective tissue and vessels, and is situated between the puncta lachrymalia into which it serves to direct the tears. Before departing from consideration of the eyelids we may notice that into the inner surface of the tarsal ligament is inserted by a broad tendon the

Levator palpebræ superioris internus, a muscle which arises from the upper part of the orbital hiatus above rectus superior, and passes within the periorbital membrane to become inserted as described. By playing over the eyeball it is enabled to raise the eyelid.

The eyelids receive blood from the ophthalmic artery, from the facial artery, and from the ocular and supra-orbital branches of the internal maxillary artery. Their nerves are derived from the ophthalmic division of the fifth cranial nerve, the supra-orbital branch of which is remarkable as passing to the foramen of that name from the foramen lacerum orbitale externally to the periorbitale.

On the supero-posterior part of the head, on either side of the poll, are situated the structures which compose the **external ear**. These consist of three cartilages on each side, of the muscles which act upon them, and of the skin which covers one of them extending inwards to line the external ear. Situated nearest to the orbit, resting on the rounded mass which is named the temporal muscle, is an irregularly triangular shaped cartilage, which seems to be connected to the other cartilages by muscle alone : this is the **scutiform cartilage**. From its external surface the external **scuto-auriculares muscles** run to the conchial cartilage, and similar muscles may be found attached to its internal surface. The **conchial cartilage** is a portion of membraniform cartilage twisted upon itself in such a manner as superiorly to present a sharp point, externally a wide broad slit, infero-posteriorly a cup-shaped diverticulum, infero-anteriorly a tubular portion along one side of which we see an irregular slit running outwards, filled with fibrous tissue (and from the other side a thin process runs to be attached below to the guttural pouch), which is connected to the external auditory process by a ring-shaped piece of cartilage, the **annular cartilage.** This cartilage is connected to the external auditory process by yellow elastic tissue, and similarly to the conchial cartilage ; the circular portion of the conchial will fit on the annular, and the annular on the bony process like the joints of a telescope, but this action is regulated by muscles which run from one cartilage to the other, which with those running from the scutiform cartilage constitute the **intrinsic** muscles of the external ear. To the conchial cartilage run the **extrinsic auricular muscles**; those anteriorly situated are the **attrahentes**, and are attached to the zygomatic process and to the parietal ridges. The **retrahentes** are attached to the ligamentum nuchæ, and by areolar tissue to the muscles situated externally to the muscles of the poll. The **adductor muscles** run from the crest of the occiput; debility of these causes the animal to become "lop-eared." **Parotideo auricularis** is the *abductor*; it runs from the external part of the base of the conchial cartilage, to become attached below to the external surface of the parotid gland. It is not directly subcutaneous, but covered by a thin layer of panniculus. Around the base of the conchial cartilage is a quantity of fat, which facilitates its movements. The common integument covering the external surface of this cartilage at its upper third is reflected over its margin at the **meatus auditorius externus (or slit)** to line the conchial and annular cartilages, and the external auditory canal, at the bottom of which it covers the external surface of membrana tympani. It is very fine and at the internal surface of the conchial cartilage presents *long hairs which seem to close up the external opening;* they serve to prevent the ingress of foreign matters. More internally it presents sebaceous glands modified for the production of the cerumen or wax of the ear, **ceruminous glands.** We notice that the external ear of the ass is more developed than that of the horse. The external ear receives blood by the **anterior and posterior auricular branches of the external carotid artery**; the latter sends a large branch, *middle auricular*, to the internal surface of the conchial cartilage. The **nerves** of the external ear are derived from *the facial, fifth cranial, and first cervical nerves.*

The inferior maxilla must now be partially removed by incision through the interdental space and through its neck after masseter externus has been separated from its inferior attachment, and thrown upwards over the zygomatic spine. By removing the piece of bone thus separated we expose the anterior attachment of digastricus inferiorly, and above this we cut through the attachment of

Mylo-hyoideus, which arises from the alveolar ridge at the superior part of the internal surface of the inferior maxilla. Its fibres run towards the centre of the intermaxillary space, where they blend with those of its fellow, thus forming a sling for the tongue and a central raphé, which posteriorly is attached to the inferior margin of the spur process of os hyoides. To the internal surface of the ramus of the inferior maxilla is attached

Masseter internus, which arises from the under surface of the crus of the sphenoid bone and of the palatine bone, and becomes attached below to the greater part of the internal surface of the ramus.

By division of its inferior attachment we expose the **inferior maxillary division of the fifth cranial nerve**, which emerges from the cranium through the anterior part of the foramen lacerum basis cranii. It sends a large bunch of fibres to join with the seventh nerve, and pass over the cheek to form the facial; just before arriving at the posterior maxillary foramen it breaks up into three parts—the *buccal, gustatory, and inferior dental branches*. The **buccal branch** winds round the front of the coracoid process of the bone and then passes over buccinator and caninus, sending large branches to the molar salivary glands. Just within its origin and firmly attached to it by nerve-fibres is the **otic ganglion**, which receives motor fibres from the seventh and sends different branches to the middle ear. The **gustatory branch** receives an extremely important branch (**chorda tympani**), which passes from the seventh nerve during its course through the middle ear. It then passes between masseter internus and pterygoideus, and goes to supply the posterior and lateral parts of the tongue with common sensation and with the special sense of taste. It also sends a very distinct branch to the sub-lingual salivary gland. The **inferior dental branch** at once passes through the posterior maxillary foramen; it then runs along below the fangs of the inferior molar teeth, between the two plates of the bone, taking a curved direction downwards and forwards; it sends a large branch to each molar tooth, and at the anterior maxillary foramen divides, one portion running onwards between the plates of the bone, to be expended in supplying the canine and three incisor teeth of that side, while the other runs through the foramen to the lower lip, blending with a few fibres from the facial nerve. The two last-mentioned nerves are separated from the buccal branch by

Pterygoideus, which arises from the under surface of the pterygoid bone and the crus of the sphenoid, and becomes inserted into the internal surface of the neck of the inferior maxillary condyle. The upper extremity of the inferior maxilla should now be removed by disarticulation of the joint, and sweeping a knife around the coracoid process in the temporal fossa. It will be thus shown that a considerable amount of fat assists in filling up this fossa, thus allowing play of the coracoid process. Deficiency of this fat causes that depression above the eyes noticeable in old or emaciated horses. In addition we also see in this fossa the

Temporalis, a muscle which passes from its attachment to the superior bony boundary of the temporal fossa, extending as far as the parietal ridges to become inserted into the inner surface of the coracoid process of the inferior maxilla. It is sometimes composed of two distinct parts, and serves when acting alternately with its fellow to give the jaw the lateral motion necessary for the grinding of the food; when acting in conjunction with its fellow it aids in closing the mouth.

Muscles which close the mouth :—Masseter externus, masseter internus, temporalis, pterygoideus. We have now exposed the posterior part of the orbit, which in the horse consists of fibrous membrane, the *periorbitale;* below this is the **internal maxillary artery**. Commencing at the termination of the external carotid, at the inner surface of the temporo-maxillary articulation, this vessel takes a direct course towards the sphenoideal foramen, and having passed through it gains foramen rotundum, and thus becomes again visible just below the orbit. It takes a straight course towards the maxillary hiatus, where it breaks up into three branches which adopt the names of the foramina of the hiatus through which they pass. Before arriving at the sphenoideal foramen this artery gives off *deep temporal branches* upwards and *branches downwards to masseter internus.* Also the **inferior maxillary artery** which passes through the posterior maxillary foramen, and courses with the corresponding nerve between the plates of the inferior maxilla supplying the lower teeth, and sending a small branch outwards through foramen menti. In its course below the orbit the internal maxillary artery gives off *branches upwards to the temporal muscle, and ocular branches,* which run to the eyeball and to the appendages of the eye, also the *supra-orbital artery*, which passes in a direction upwards to the root of the orbital process of the frontal bone; after passing through the supra-orbital foramen it breaks up in supplying the muscles of the forehead. A remarkable branch of this artery, the **lateral nasal** makes a peculiar twist before piercing the internal orbital foramen. By passing through this *it is described as gaining*the nasal chamber, from which it passes through the cribriform plate of the ethmoid bone into the olfactory sinus, and after supplying the olfactory bulb it passes again through the cribriform plate and supplies the mucous membrane of the superior meatus of the nasal chamber. Some of the ocular branches passing over the lower margin of the orbit anastomose with branches of the facial artery. The internal maxillary artery sends branches downward to supply the soft palate, and the buccal branch to the cheek and the molar glands; it then breaks up into the spheno-palatine, palato-maxillary, and superior dental branches.

The **spheno-palatine artery**, passing through the foramen of the same name, gains the nasal chamber, where it supplies *the inferior and middle meatus*. The **palato-maxillary artery**, on emerging from the palatine canal through the palatine foramen, courses its way along the lateral part of the hard palate, and bending inwards anteriorly is retained in position by a small cartilaginous hook, and anastomoses with its fellow, the arch so formed giving off a branch which passes through foramen incisivum to anastomose with the terminal parts of the superior labial arteries. The **superior dental artery** passes through canalis infra orbitale, and sends a branch to each molar; it sends a small branch through the infra-orbital foramen to anastomose with the arteries of the face, but the main trunk runs between the plates of the bones to supply the tush and the incisors.

The **fifth cranial nerve, pars trigemini,** arises by two roots from the lateral part of the pons Varolii at the base of the brain. One of these is sensory, the other is motor; in this respect the nerve resembles those given off from the spinal cord; it is hence sometimes called the spinal nerve of the brain, and the analogy is increased by the presence of a large ganglion, the **Gasserian,** on the sensory root. This ganglion is situated between the two folds of dura mater, which constitute the membranous tentorium, and the motor root runs over its under surface. From the ganglion three branches pass:—One runs to combine with the motor tract to form the *inferior maxillary division*, which passes through foramen lacerum basis cranii, as already noticed; another, the *ophthalmic division*, passes through foramen lacerum orbitale, in company with the third, fourth, and sixth cranial nerves and the ophthalmic artery, and thus gains the orbit. The third, or **superior maxillary division**, passes from the cranium, and emerges with the internal maxillary artery through foramen rotundum. It may be seen with this vessel passing to the maxillary hiatus, prior to gaining which it divides into three parts, corresponding to the terminal branches of the artery. The *superior dental branch* passes through canalis infra-orbitale, and, like the artery, supplies the upper teeth; but a very large branch passes through the infra-orbital foramen, where it is covered by levator labii superioris alæque nasi. It blends with a large branch from the facial nerve, forming the superior labial nerve, and *the plexus of nerves over the lateral part of the muzzle formed by the union of the infra-orbital, anterior maxillary, and facial nerves corresponds to the* "**pes anserinus**" **of the human subject** *The palato-maxillary nerve*, passing with the artery of the same through the palatine canal, gains the hard palate, which it supplies with sensation; it also sends fibres to the soft palate. The *spheno-palatine nerve*, passing with the spheno-palatine artery through the foramen of the same name, gains the nasal chamber, the *inferior and middle meatus* of which it supplies with common sensation. At its origin it forms a close connection by means of numerous fibres with the **spheno-palatine or Meckel's ganglion,** a collection of nerve-cells, which *receives motor fibres through the* **vidian branch** *of the seventh cranial nerve*, and is supposed to send efferent fibres to the palate.

We have now exposed the **mouth,** a cylindrical cavity elongated in an antero-posterior direction, presenting four surfaces and two extremities. The *anterior extremity* is guarded by the lips, between which is the *labial fissure*. The *posterior extremity* presents the anterior opening of the fauces, bounded on either side by the reflections of mucous membrane from the palate to the base of the tongue, the *posterior pillars of the tongue*. Through this opening the pendulous palate and the cavities of the tonsils are visible.

The *lateral surfaces* of the mouth are formed by the cheeks, the mucous membrane lining these presents perforated papillæ, on which the ducts of the buccal glands pour the secretions of those glands; varying in position in different subjects, but generally about opposite the space between the second and third molar teeth, we may distinguish the opening of the parotid duct with a larger papilla. The mucous membrane posteriorly is reflected inwards behind the molar teeth to join the posterior pillars of the tongue; in some places it is coloured by pigmentary deposit. On the roof of the mouth, centrally, it is thick, and presents dense stratified epithelium and a plentiful submucous tissue, in which is situated a remarkable plexus of veins congestion of which from dental or other irritation causes that prominence of this part termed "lampas" by individuals who ignorantly considered it a primary diseased condition; on either side of this runs a palatine artery, as has been described. That portion of the mucous membrane which is attached to the bony palate is termed the **hard palate**, and inferiorly presents a central groove, from which crescentic prominences, seventeen to eighteen in number on each side, run in an oblique direction outwards. These are termed the "**bars**," and are straightest and closest posteriorly, where they blend with the

Soft palate, the mucous membrane of which is redder and less dense, and which runs in a backward direction from the posterior crescentic margin of the palatine bones.

Occupying the outer circumferent margin of the roof and of the floor of the mouth at particular parts we see the **teeth**. These are products of mucous membrane. They are hard organs; by some anatomists considered as bones, but are merely the result of special development of mucous membrane. They are firmly fixed in cavities in bone termed *alveolar cavities*, and are surrounded by mucous membrane termed the gums. They present three parts:—
A *fang*, that part enclosed in the alveolar cavity; *crown*, that visible on inspection of the mouth after the teeth are fully developed; *neck*, that part connecting the fang and the crown, and to which the mucous membrane forming the gum is attached. Teeth are composed of three substances (two of which closely resemble bone in histological structure), dentine, enamel, and crusta petrosa or cement. They surround a central cavity, which is called

the pulp cavity, for it contains a small papilla to which an artery and a nerve run and from which a vein proceeds ; this is the **pulp**. It is immediately surrounded by **dentine**, which is formed continuously from its outer part, so that the pulp cavity, which is large in the young animal, in the adult becomes almost obliterated ; on the outer surface of the dentine rests a layer of very dense white substance of a petrous nature, **enamel**. It is deficient on the lower part of the fang. Externally placed to this we see a yellowish-brown matter, **crusta petrosa or cementum**.* The upper surface of the tooth generally presents one or more depressions, the **infundibuli**, into which dip the three layers of substance of which the tooth is composed, and which are frequently filled with crusta petrosa. When a tooth is wholly covered on its external surface with enamel it is termed a *simple tooth*, all other teeth are *compound*. We see simple teeth in the dog ; the canines of the horse are by some considered simple. Teeth are named according to their function, *grinders and pincers*. These are arranged in rows, each of which consists of six teeth, and between the rows of grinders and pincers in the horse are situated solitary teeth, the **canines, cuspidati, or tushes**, which in the mare are rudimentary ; they are weapons of offence or of defence, and are well developed in stallions. The **incisors or pincers** are situated in the mouth, anteriorly in contact with the inner surface of the lips, arranged along the anterior crescentic margin of the hard palate ; a corresponding number being placed in the upper margin of the lower jaw, three incisors on each side of the symphysis. The upper incisors are larger, longer, and put in their appearance later than those of the lower jaw. The cutting surface of the teeth is ovoid, with an acute angle at either extremity ; the anterior edge is higher than the posterior, and the tooth, taken as a whole, presents a curve the concavity of which is posteriorly situated. There are two complete sets of incisors, **temporary or milk**, and perma-

* Crusta petrosa in structure exactly resembles bone presenting laminæ, lacunæ, and canaliculi, and, where it is very thick, Haversian canals. Dentine consists of an earthy matrix, through which branching and wavy canaliculi run from the pulp cavity, and near its outer surface terminate in lacunæ. Enamel consists of prismatic hexagonal bodies, which have their margins jagged to enable them to fit closely together. One extremity of these looks towards the pulp cavity, the other towards the surface of the tooth.

nent. These are distinguishable by the following characteristics:—Milk teeth are smaller and whiter than permanent, and their infundibulum is more shallow; they have a marked neck, and are smooth on the anterior surface of the crown; permanent incisors decrease in size gradually from the nipping surface to the extremity of the fang, and present grooves on their anterior surface, running in a direction from above downwards. At an early period of foetal life there may be distinguished the smooth margin of the jaw covered by mucous membrane. A longitudinal groove shortly presents itself, and at the bottom of this "*primitive dental groove*" mucous papillæ appear. These are produced by a thickening of the mucous corium, which causes an elevation of the epithelial and basement layers. Transverse partitions (which subsequently become converted into bone) are thrown across the groove, whereby it is separated into as many cavities as it subsequently produces teeth (*follicular stage*). Then by gradual closure of the upper part of the groove, by approximation of the margins, each papilla becomes enclosed in a sac (*saccular stage*), which is lined by epithelial cells. The parietal portion of the epithelial layer is said to produce the crusta petrosa, the visceral (that clothing the papilla) epithelium is converted into enamel, and the corium is continuously developed into dentine, that part which has not yet been changed being termed the dental pulp. The papilla thus altered becomes too large for its cavity and forces its way through the gum which forms the upper wall of the sac (*eruptive stage or "cutting the teeth"*). Small portions of the dental sac are separated from the main cavity at an early stage of the developmental process, in those cases where a second or permanent set of teeth will be required to replace the temporary organs. These small sacs are termed *cavities of reserve*, and slowly undergo the changes above described in connection with the parent sac, a papilla being produced which subsequently becomes developed into a permanent tooth, by the growth of which absorption of the fangs of the temporary teeth is brought about, and they, being deprived of their source of nutriment, fall from the mouth ("*shedding the milk teeth*"). Of the molars, the three anterior of each row only are replaced, for in the jaw of the young animal there is not room enough for six molars. Those teeth which occupy the three posterior

SPECIAL ANATOMY.

alveolar cavities are developed late, and remain throughout life, the sixth or posterior one not appearing until the three anterior (temporary) molars have been replaced, as the following table indicates—

	Temporary.	Permanent.
1st incisor	At birth	2 years 6 months.
2nd „	6 weeks	3 „ 6 „
3rd „	9 months	4 „ 6 „
Tushes	—	4 „ 6 „
1st molar	At birth	2 „ 6 „
2nd „	„	2 „ 9 „
3rd „	„	3 „ 6 „
4th „	—	1 year 3 „
5th „	—	1 „ 9 „
6th „	—	4 years 6 „

Variations, however, occur as a result of breed, management, &c. *Thoroughbred horses date their birth from the 1st January, all other breeds from the 1st May.*

The **molars** are developed from compound papillæ. They are found in four rows of six each, one on either side of the upper and lower jaws. Each molar presents six surfaces:—The *anterior* is smooth, in contact with the tooth in front through the medium of the thin plate, forming the anterior boundary of the alveolar cavity. But in the first molar this margin is sharp. The *posterior surface* is in similar contact with the tooth behind, but in the last molar is sharp. The *external surface* presents two grooves running from above downward, of which *the anterior is the deepest;* the *internal surface* is also grooved, but less distinctly, *the outer surface of the upper, the inner surface of the lower, molar is the largest,* for the *superficial or grinding surface* in both the upper and lower molars slants obliquely downwards and outwards, so that the inner part of the upper comes in contact with the outer part of the lower molars, a consequence of the greater width of the superior than of the inferior maxilla; so in rasping a horse's teeth the rasp must be applied to the inner edge of the lower, the outer edge of the upper, grinders. The grinding surface presents the edges of the layers of the substances of which the tooth is composed arranged " in the form of a Gothic B," so that, as substances of different degrees of durability are exposed to equal attrition, an irregular surface favorable to the comminution of the food is constantly maintained. The superior molars present a square, the inferior an oblong grinding surface. The *deep-seated surface* presents deep pulp cavities into which the portions of the divided pulp fit.

In some cases **premolar** or "*wolves teeth*" may be found fixed in the jaw in contact with the anterior surface of the first molar. They are small portions of tooth substance, generally forced out by growth of the permanent molars, which occupy more room than the temporary. The **canines, tushes, or cuspidati** are four in number in the male, and occupy the space between the molars and incisors, being about equidistant from both. The inferior canines are more anteriorly situated than the superior. These teeth present no infundibula, but have a single fang, and their crown terminates in a point at its free extremity, at which two ridges which separate it into two surfaces terminate. The external surface is convex in every direction; the internal surface presents a groove surrounding a prominence. This is the condition of the unworn tooth; in the mouth of the old animal the crown becomes conical and blunted. In the mare these teeth may be sometimes seen just emerging from the gum, or simply as a papilla changed into tooth substance on the surface of the mucous membrane.

The *floor of the mouth* therefore anteriorly and on either side presents teeth, and the mucous membrane from the lower margin of the cheeks and lower lips is reflected upwards to form the gum, and from the teeth on the inner side after passing for a short distance downwards it is reflected upwards to cover the tongue, which occupies the central part of the floor of the mouth. In doing so it forms a marked fold at the central line anteriorly, which serves to prevent too free play of the tip of the tongue, running upwards from against the symphysis of the lower jaw and enclosing a portion of the two genio-hyoglossi muscles. This is the **anterior pillar or frænum linguæ**; on either side of it may be distinguished an opening of a submaxillary salivary '(Wharton's) duct, guarded by a papilla termed the "**barb**."

The **tongue** is the organ of the special sense of taste, and is prismatic and thick posteriorly, and terminates in a rounded spathulate extremity anteriorly. Its bulk consists mainly of muscular structure mixed with the fibres of which is much fat. Against the superior surface the fatty and muscular structures become intimately blended with the deep surface of the mucous membrane, thus constituting a somewhat firm layer into which some of the more definite muscles are inserted. The mucous membrane of the lateral part of the tongue presents several small papillæ on which *the ducts of the sublingual salivary glands* open. These may be seen as thin white threads running from the glands, on elevation of the mucous membrane. Superiorly to these we see somewhat large **fungiform papillæ**, so named from their resemblance to a mushroom in shape. These are found only on the lateral parts of the tongue, and on their surface present **filiform papillæ**, larger specimens of which may be found on the major part of the *dorsum or upper surface of the tongue* to which they communicate the peculiar velvety feel. They are hair-like prolongations of the mucous membrane, and are either *simple or compound*. The anterior part of the dorsum of the tongue presents a central* fissure, the result of the frequent longitudinal folding of this portion of the organ. The anterior edge of this part is rounded and the inferior surface along the central line presents the attachment of the frænum. The posterior part of the dorsum surface of the tongue, elevated above the anterior part, presents two remarkable round spots which, on examination, prove to be collections of fungiform papillæ surrounded by a deep groove (**circumvallate papillæ**). These are situated centrally about an inch from each other. A smaller one may be sometimes seen just behind them.

We must now examine the **hyoid bone or bone of the tongue**. This forms the basis of attachment of the pharynx, larynx, and other structures situated in this neighbourhood. It is suspended from the under surface of the cranium, and hangs between the rami of the inferior maxilla. It consists of five distinct portions: a *main portion* and *four cornua*. The *main portion* closely resembles a hunting spur, being composed of a **spur process**, running in an anterior direction, and two **heel processes**, which diverge from each other and run backwards about three inches, "forming a crescentic space between them." These two processes terminate posteriorly each in a point which is continuous with the free extremity of the cornu of the thyroid cartilage. They present a superior sharpened margin for muscular attachment, and an inferior margin to which the thyro-hyoid ligament and muscle are attached. To their outer surface posteriorly is attached hyo-thyro-pharyngeus a constrictor of the pharynx. The spur and heel processes are connected together by the **body** of the bone, which presents two surfaces and four borders. To the *posterior border* **hyo-epiglottideus** runs from the base of the epiglottis, and the thyro-hyoid ligament is attached. From the *anterior border* the spur process projects. The *lateral borders* posteriorly give rise to the heel processes, anteriorly present circular synovial articulatory surfaces which look in a forward, upward, and outward direction; to these the inferior extremities of the short cornua are attached. The *superior surface* of the body and the upper sharpened border of the spur process afford attachment to **retractor lingualis**, an obscure muscle, the fibres of which, imbedded in the fatty substance of the tongue, influence its movements. The *inferior surface* of the body receives the superior terminal portions of subscapulo-hyoideus and sterno-thyro-hyoideus. The inferior sharpened border of the spur process affords attachment to mylo-hyoideus and its extremity, which is its deepest part, to

Genio hyo-glossus, the muscle which lies in contact with its fellow at the central plane of the tongue, and which, being attached to this point and to the symphysis of the inferior maxilla, sends fibres in a radiating direction upwards to become lost in the musculo-adipose matter of the tongue as far forward as its tip.

Genio-hyoideus also runs from the anterior extremity of the spur process of the os hyoides. It becomes attached anteriorly to the symphysis of the jaw, and lies in contact with its fellow in the central line of the inter-maxillary space.

Extending in an oblique direction forwards and upwards from the synovial articulatory surfaces of the lateral parts of the body of the hyoid bone are the **short cornua**, cylindrical portions of bone shorter than the heel processes, smallest at the centre, superiorly connected by fibro-cartilage (*in which is generally to be found a bony nodule analogous to the middle cornu of the ox*) to the inferior extremity of the long cornu. These are embedded in the base of the tongue and are connected together by the

Hyoideus transversus, which runs from the internal surface of one short cornu to that of the other.

Hyoglossus brevis, the fibres of this muscle arise from

the whole length of the lateral part of the main portion of the hyoid bone, and radiate from this toward the under surface of the mucous membrane of the base of the tongue. Under this muscle, and so closely blended with it as to be distinguished only by the direction of its fibres which run in a parallel direction forwards, is

Hyoglossus parvus, arising from the junction of the long and short cornua of the hyoid bone, and blending with hyoglossi magnus and brevis in the substance of the tongue. To the posterior margin of the short cornua and of the inferior extremity of the long cornu the

Hyoideus parvus is attached, and runs obliquely downwards and backwards to the superior margin of the heel process.

The **long cornu** (of which there is one on each side) runs obliquely upwards and backwards from the superior extremity of the short cornu. It presents three angles, three borders, and two surfaces. It assumes the form of an extremely acute angled isosceles triangle, with its apex downwards and forwards. The *inferior angle* is connected by fibro-cartilage with the short cornu, the *antero-superior angle* is connected by fibro-cartilage with the hyoid process of the petrous temporal bone. From the *postero-superior angle*, which is somewhat tuberous, the hyoideus magnus and digastricus (sometimes) arise.

Hyoideus magnus becomes inserted into the supero-lateral part of the body of os hyoides by a small tendon, the fibres of which separate, the foramen thus formed being lubricated with synovia for the passage of the tendon of

Digastricus, which commences generally from the styloid process of the occiput, being blended with the origin of stylo-maxillaris, forms a fusiform belly, and then a thin round tendon which plays through that of hyoideus magnus, and terminates in the second muscular portion which runs forwards, becoming inserted into the inferior margin of the lower jaw as far forward as the symphysis. The concavity of this muscle therefore is anteriorly situated. The *anterior margin* of the long cornu presents a deep curve superiorly, where the mucous membrane of the guttural pouch is reflected from the internal surface to the external surface of the bone. To the lower part of the *internal surface*, however, hyo-pharyngeus becomes attached, and from a ridge running across the lower part of the *external surface*, a long fusiform muscle—

Hyoglossus longus—runs forwards along the lateral part of the tongue, immediately beneath the mucous membrane.

The superior margin is the shortest, and affords attachment to

Stylo-hyoideus, which arises from the styloid process of the occiput, and becomes inserted into the superior margin, and external surface of the extreme upper part of the long cornu.

The *posterior margin* is almost straight, between it and hyoideus magnus the external carotid artery passes superiorly; and lower down the twelfth or lingual nerve, and the submaxillary artery, which on gaining the external surface of the long cornu gives off the

Lingual artery, which runs in a forward direction over the external surface of the short cornu, through mylo-hyoideus, the fibres of which separate for its passage, and breaks up to form the ranine and sublingual branches. The **ranine** runs to the tip of the tongue between the genio-hyoglossus and genio-hyoideus muscles, taking a wavy course, whereby it is capable of adapting itself to temporary elongation of the tongue. It sends branches downwards to anastomose with the superior branches of the **sublingual**, which, after supplying the gland of the same name, expends itself in the muscular structure of this neighbourhood. The **veins of the tongue** correspond to the arteries, and empty themselves into the submaxillary vein. The **nerves** are the *gustatory branch of the inferior maxillary division of the fifth*, and the *anterior branch of the glosso-pharyngeal (ninth), which endow it with common sensation, and with the special sense of taste. It owes its motor power to the* **twelfth, lingual, or hypoglossal nerve**, which arises from the lateral part of the medulla oblongata, and emerges from the cranium, through foramen condyloideum. It passes over the guttural pouch, and runs in an anterior direction between the pneumogastric and spinal accessory nerves, over the internal and external carotid arteries, and between hyoideus magnus and the long cornu to the lateral part of the tongue.

By division of the amphiarthrotic joint of the hyoid long cornu with the petrous temporal bone, and bending downwards this process, we expose the **guttural pouch.** This is a dilated pouch, lined by a transparent mucous membrane in which we may see vessels ramifying, and which has an opening anteriorly situated, a long slit in the wall of the Eustachian tube, which here degenerates into a mere groove between two cartilaginous prominences, dilated inferiorly. It is visible as far as foramen styloideum, and inferiorly is the only source of communication between the pharynx and the guttural pouch. The outer and lower corner of the anterior surface of the pouch is bounded by hyo-pharyngeus, situated between stylopharyngeus and palato-pharyngeus; here also we see the **parotid lymphatic glands.** Posteriorly the guttural pouch presents the superior attachment of the recti capitis antici muscles, on the outer side of which are large nerves and the internal carotid artery. The pouch along the central line lies in contact with its fellow. It presents numerous papillæ on its internal surface. By cutting through the posterior pillar of the tongue we now expose the **isthmus of the fauces.** This is the passage which runs from the base of the tongue to the epiglottis. Anteriorly it is bounded by the opening between the posterior pillars of the tongue, posteriorly by the opening, on either side of which are the **pillars of the fauces**, which are composed of muscular fibres running directly from the pterygoid process to the thyroid cornu, surrounded by mucous membrane, much thickened by addition of salivary glandular structure. The roof of this passage is formed by the **soft palate,** which consists of a fixed and pendulous portion. The *fixed portion* is attached anteriorly and laterally to the crescentic posterior margin of the ossa palati. It consists of a fibrous layer, the superior surface of which is covered by pharyngeal mucous membrane, the inferior separated by a muscular layer and salivary glandular structure from the mucous membrane of the fauces. This muscular layer, **palato-pharyngeus,** runs from the fibrous layer and from the pterygoid process, to blend with its fellow at the central line of the pharynx, and to become attached to the thyroid cornua. The fibrous layer is also acted upon by

Tensor palati, which arises from the styloid process of the petrous temporal bone in common with stylopharyngeus, runs forwards along the side of the Eustachian tube, to which it is also attached, and after passing through the pterygoid groove, where it is bound down by a ligament, and lubricated by synovia, becomes inserted into the fibrous palate, which it serves to render tense.

The **pendulous portion** of the soft palate is very large in the horse, and serves to close the posterior opening of the fauces, except during the passage of food. Its inferior margin is somewhat concave, and it consists simply of two layers of mucous membrane, with an intervening glandular layer. The *floor of the fauces* is that portion of mucous membrane situated between the base of the tongue and the epiglottis. Every portion of the mucous membrane of the fauces presents large papillæ and openings, for it is surrounded by the **palatine salivary glands**. Its *lateral walls* present depressions corresponding to the prominences termed tonsils in the human subject. The **pharynx** is an elongated passage of mucous membrane, into which several muscles are inserted, and on which they act. Superiorly the pharynx presents anteriorly the *two posterior openings of the nasal chambers*, separated by the inferior sharpened margin of the vomer, closely invested by mucous membrane; on either side of these, and more posteriorly, are elongated slit-like openings, each guarded by a cartilaginous valve, the *openings of the Eustachian tubes*, through which air passes to the middle ears and to the guttural pouches, more posteriorly the muscular walls of the pharynx lie in contact superiorly with the guttural pouches. Posteriorly the *commencement of the esophagus* is the result of the gradual narrowing of the cavity. The floor of the pharynx is formed anteriorly by the upper surface of the soft palate, posteriorly by the *superior orifice of the larynx*; between the two is the *posterior opening of the fauces*. The **muscles of the pharynx** are either constrictors or dilators. **Constrictors**: *Hyo-thyro pharyngeus*, runs from the posterior part of the heel process of os hyoides and from the thyroid cornu. *Thyro-pharyngeus*, from the lateral part of the external surface of the ala of the thyroid cartilage. *Crico pharyngeus*, from the lateral part of the cricoid cartilage (this muscle is directly continuous with the muscular layer of the esophagus). These three muscles run upwards to blend with their fellows and with each other in becoming inserted into the central line of the upper surface of the pharyngeal mucous membrane. **Dilators**: *Hyo-pharyngeus*, runs from the internal surface of the long cornu of os hyoides, and blends with its fellow upon the upper surface of the pharynx. *Stylo-pharyngeus* arises in common with tensor palati from the styloid process of the petrous temporal bone, and also is attached to the lateral part of the Eustachian tube; it blends with its fellow on the wall of the pharynx. *Palato pharyngeus* has been already noticed.

The pharynx is supplied with blood by the *ascending pharyngeal branches of the submaxillary artery*, and also by *branches from the internal maxillary*. It is partly supplied with nerve force by the *sympathetic*, partly by the *glosso-pharyngeal, and superior laryngeal branch of the pneumogastric*. The **glosso-pharyngeal (or ninth) cranial nerve** arises from the lateral part of the medulla oblongata, and in passing out of the cranium by foramen lacerum basis cranii has a ganglion upon it. Passing over the wall of the guttural pouch it has intimate connection with the large nerves of this neighbourhood and with the superior cervical ganglion. It then divides, sending one branch to supply the posterior and lateral parts of the tongue with common sensation and the special sense of taste; the other to the pharynx.

By an incision in a longitudinal direction through the face to one side of the central line we may expose the nasal chambers, and by removal of an oblong portion of the septum nasi we may expose the turbinated bones *in sitû*. The **NASAL CHAMBERS** are cavities elongated in a direction from before backwards, bounded *anteriorly* by the anterior nares, *posteriorly* by the posterior nares. The chamber of one side is separated from its fellow inwardly by the **septum nasi**, a thick layer of cartilage, which is attached along its inferior margin into the groove of the vomer; posteriorly to the crista galli process of the ethmoid bone; superiorly it bifurcates and becomes spread out over the lower surface of the nasal bones, beyond which it extends anteriorly, forming a point in front of the nasal peak, to which, on either side, an inverted comma-shaped piece of cartilage is attached, being composed of the alal and corneal cartilages. The **alal cartilage** is circular, presenting a superior and an inferior surface, and from its inner margin the **corneal cartilage** runs in a direction forwards, downwards, and outwards. The inferior extremity of the latter is continuous by means of a fold of mucous membrane with the inferior turbinated bone.

Dilator naris anterior runs from the external surface of one ala to that of the other, while the corneal cartilages are connected together anteriorly by a direct continuation of it, by some anatomists termed **cornealis transversus**. Dilatores naris superior and inferior, levator labii superioris alæque nasi, and retractor labii superioris, also become attached to these cartilages. The skin extends around the margin of the nostril, as formed by these structures, and is continued for a short distance in the nasal chamber, where it insensibly blends with the mucous membrane. About the point of junction at the outer side, inclined to the superior part of the nostril (infero internal in the *horse*), is the *anterior opening of the ductus ad nasum* which looks as though "punched out of the mucous membrane." On the outer side the common integument becomes very thin, hairless, and prolonged in the angle between the superior process of the anterior maxilla and the free margin of the nasal bones, to form a blind pouch, the **false nostril**, which we have found filled with epithelial layers of cells. To the posterior extremity of this a muscle which runs from just above the anterior termination of the zygomatic ridge is attached, also levator labii superioris alæque nasi. Into this cavity the air passes directly from the anterior naris; it has to diverge slightly inwards to gain the nasal chamber. The *roof of the nasal chamber* is formed by the inferior concave surface of the nasal bone, posteriorly by the nasal plate of the frontal bone. The *outer wall* presents the two ossa turbinata, by means of which the cavity is divided into three meatus.

The **superior turbinated bone** is attached to the inner surface of the nasal bone, and consists of a thin plate curled from above downwards. Anteriorly it terminates by a cartilaginous prolongation, which gradually blends with the wall of the chamber; posteriorly it is directly continuous with the lateral mass of the ethmoid bone. Its *cavity* is divided by a partition into two parts; the anterior helps to form the nasal chamber opening into the middle meatus; the posterior forms part of the superior maxillary sinus.

The **inferior turbinated bone** is attached to the inner surface of the superior maxillary nasal plate. It is smaller than the superior; posteriorly it gradually disappears; anteriorly it presents a bifid cartilaginous prolongation, one of the divisions of which runs to the inner, the other to the outer wing of the nostril. It has also an anterior and a posterior division; the former helps to form the middle meatus, the latter enters into the composition of the inferior maxillary sinus. Thus the **superior meatus**, the longest and narrowest, is between the roof of the chamber and the upper surface of the superior turbinated bone. It extends from the anterior naris to the cribriform plate. The **middle meatus**, between the ossa turbinata, comprises the anterior cavities of both turbinated bones, and extends from the anterior naris to the ethmoid cells, blending anteriorly with the superior; *posteriorly it presents the opening of the sinuses of the head.* The **inferior meatus** is between the lower surface of the inferior turbinated bone and the floor of the chamber; it extends from the anterior to the posterior naris, blending anteriorly and posteriorly with the middle. The **Schneiderian or pituitary mucous membrane** lines every part of the nasal chamber. It is highly vascular and thick, performs the part of periosteum to the turbinated bones and ethmoid lateral masses, and assumes almost an erectile character at the anterior extremity of the ossa turbinata. The superior meatus is supplied by the lateral nasal artery and nerve, and the inferior and middle meatus by the spheno-palatine artery and nerve. The ethmoid cells and the posterior part of the nasal chamber, in addition, receive the fibres of the **olfactory nerve**, which afford the special sense of smell. These pass through the foramina in the cribriform plate of the ethmoid bone, and are not enclosed in any common sheath, but each fibre runs its course independently after its origin from the olfactory bulb, and terminates in a cell on the nasal mucous membrane. These fibres, examined under the microscope, are found to resemble sympathetic fibres, being devoid of medullary sheath and of white substance of Schwann. The septum nasi, at its antero-inferior part, is continuous with the cartilage which fills the incisive openings, and in which may be found a mucous organ resembling **the duct of Stenson**, which in the ox forms a communication between the nasal chamber and the cavity of the mouth. In the horse it only opens into the nasal chamber. We have traced it backwards above the bony palate, as far as the soft palate, where it terminated in a cul-de-sac. Some of the bones of the head present large cavities lined by mucous membrane, which is (normally) thin and transparent. These, the

Sinuses of the head, are five in number on each side. They present numerous bony trabeculæ projecting inwards forming partial septa, and are best developed in old animals, being the result of absorption of bone. By containing rarefied air they serve to lessen the weight of the head, and by causing bulging of the bone they afford increased surface for muscular attachment. The **frontal sinus** in the bone of the same name is bounded by its cranial, nasal, and external plates. Internally it is separated from its fellow by a bony plate. To *open this sinus* we draw a line from the lower margin of the orbital process of one frontal bone to that of the other; bisect this line and in the angle so formed inferiorly trephine. The **sphenoid sinus** (which only appears very late in life) **and the ethmoid sinus** are cavities mainly contained in the bodies of the bones of the same names respectively. The **antrum or superior maxillary sinus** is formed by the palatine, superior maxillary, malar, lachrymal, and superior turbinated bones. It presents two bony canals running through it—one, canalis infra-orbitale, through which the superior dental artery and nerve run; the other is the ductus ad nasum, which terminates by piercing the nasal plate of the superior maxilla several inches up the nasal chamber; being continued by a membranous duct which runs underneath the mucous membrane to its opening. The above sinuses open by a common foramen into the nasal chamber, by a very small opening at the posterior part of the middle meatus. The **inferior maxillary sinus** is more anteriorly placed, and is separated from the superior maxillary sinus by a thin but imperforate plate which extends to the opening into the nasal chamber, thus dividing it into two. So this sinus does not communicate with the rest, all of which communicate freely among themselves. To trephine this sinus, a line is drawn parallel to the zygomatic ridge about an inch above it, and the point where this meets, a perpendicular line connecting the two from the termination of the ridge, indicates the point of incision.

The contents of the orbit may be examined by removal of the superior and outer walls by sawing through the rest of the orbital process of the frontal bone, externally to the supra-orbital foramen, and through the lower part of the malar bone. After which the separated portion of bone must be raised by a chisel, and removed by division of the peri-orbitale. We are thus enabled to examine the **lachrymal gland**, a racemose glandular body flattened from above downwards, and moulded between the under surface of the orbital process and the structures beneath; it has also an accessory lobule found in the upper eyelid. It resembles one of the salivary glands in appearance, and pours its secretion through *several small ducts* upon the inner surface of the upper eyelid. After passing over the front of the eyeball, being diffused by the winking action of the eyelids, the tears (or lachrymal secretion) pass through the canal formed by approximation of their free margins towards the inner canthus, to which it is directed by the downward and inward inclination of the palpebral fissure. By the caruncula lachrymalis it is directed into the **puncta lachrymalia**, openings situated one at the inner extremity of each eyelid, and from this passes through a series of membranous canals to the anterior naris. It first gains the **lachrymal canals**, one of which runs from each punctum in a downward direction; they converge and just at the lachrymal foramen meet and form the **lachrymal sac**, from which the **ductus ad nasum** runs through a special canal in the lachrymal and superior maxillary bones. It may be exposed by use of the hammer and chisel and will be found to extend for about three quarters of its course in the bony canal. Its anterior part is continued onwards between the bone and the Schneiderian membrane as far as the anterior naris, where it opens at the point of junction of the skin and mucous membrane at the superior part. Situated between the caruncula lachrymalis and the eyeball is the **cartilago-nictitans (winking membrane, or third eyelid)**. It is thickest posteriorly, and at its anterior free margin is covered by mucous membrane, which generally contains some pigment. Internally its yellow fibro-cartilaginous substance is continuous with some yellow elastic tissue, enclosed in the meshes of which, and running among the muscles of the eye, is **fatty matter** on *which the movements of the cartilage depend*, for when, by the action of its muscles, the eyeball is retracted it presses upon this mass of fat, which in its turn presses the cartilage over the front of the eyeball.

144   OUTLINES OF EQUINE ANATOMY.

This fat is found even in the most emaciated subjects. We have already noticed the **levator palpebræ superioris internus**; by incision through the periorbitale we may now expose it: a thin layer of muscle which arises from the upper margin of the sphenoideal hiatus and runs to the upper eyelid, on which it is enabled to act by gaining leverage over the upper surface of the eyeball. The **periorbitale** is a dense fibrous membrane which lines the orbital cavity, forming the sole protection to the contained structures at the postero-inferior and superior parts, which in the human subject are bony. Anteriorly it is attached to the margin of the orbit, posteriorly around the sphenoideal hiatus; it encloses the eyeball and its muscles. From around this hiatus other muscles arise and run to the eyeball.

Retractor oculi, consisting of four distinct portions, arises from directly around it, and after passing in a forward direction in contact with the optic nerve becomes inserted into the posterior part of the external surface of the sclerotic coat. The four

RECTI—externus or abductor, internus or adductor, superior or levator, and **inferior or depressor**—arise from the external, internal, superior, and inferior parts of the sphenoideal hiatus respectively, and pass to the anterior part of the sclerotic, and form aponeurotic expansions which unite at their margins respectively, forming thus a continuous layer, situated around the outer margin of the cornea, covered anteriorly by conjunctiva, **tunica albuginea** (forming the "white of the eye").

Superior oblique muscle arises from just above the straight muscles at the upper part of the sphenoideal hiatus. It then runs to the root of the orbital process of the frontal bone, where it passes through a fibro-cartilaginous loop. From this it runs to the upper part of the eyeball, forms an aponeurotic tendon which runs beneath the tendon of levator oculi, and becomes inserted into the anterior part of the sclerotic between the last-mentioned muscle and abductor oculi. *Strangeways has noticed an accessory muscle which arises near the fibrous loop and becomes blended with this; we have frequently verified this observation.*

Inferior oblique muscle arises from the orbital plate of the lachrymal bone, near the lachrymal foramen, runs outwards and becomes aponeurotic, is inserted into the anterior part of the sclerotic between the abductor and depressor; it passes externally to the depressor.

The structures contained within the orbit receive blood through the *ocular branches of the internal maxillary artery* and through the *ophthalmic artery*. The former we have already noticed, and of these the most remarkable, which we distinguish are the *lateral nasal*, taking a peculiar curve through the bottom of the orbit, and then passing into foramen orbitale internum, and the *supra-orbital*, which runs directly to the foramen of the same name. *Ciliary arteries* run from the internal maxillary to the eyeball. The **ophthalmic artery** arises in the cranial cavity from the anterior communicating or from the middle cerebral artery. It passes through foramen lacerum orbitale in company with several nerves, and on emerging from the foramen gives off *branches to the lachrymal gland*, and *to the eyelids*, also **arteria centralis retinæ**, which gains the eyeball by passing through the centre of the optic nerve. Three nerves arising from the base of the brain (nominally) supply the muscles of the eyeball with motor power. **Motores oculorum or third cranial pair of nerves**, arising from the centre of the under surface of the crus cerebrum, each nerve passes from the cranium through foramen lacerum orbitale and then gives off fibres to all the muscles except the abductor and the superior oblique (and the external part of retractor). From this nerve also a branch passes to the **ophthalmic or lenticular ganglion**, a small collection of nerve cells found at the point where the third nerve gives off fibres to the inferior oblique muscle. It receives sensory fibres from the lateral nasal branch of the ophthalmic division of the fifth, and sends ciliary branches to the eyeball. The **ophthalmic division of the fifth cranial nerve**, passing from the gasserian ganglion, emerges from the cranium through foramen lacerum orbitale, it then breaks up to supply the structures of the eye and its appendages with common sensation. Among others, giving off the *supra-orbital and lateral nasal nerves*, which follow the course of the arteries of the same name. *The lateral nasal affords the sensory root of the ophthalmic ganglion*. The **pathetic or fourth nerve** arises from the valve of Vieussens, which forms the anterior part of the roof of the fourth ventricle of the brain. These are the smallest nerves at the base of the brain, and wind round the outer surface of the crura cerebri, putting in their appearance just in front of the pons Varolii. They generally are found emerging from the cranium through foramen lacerum orbitale (and not through foramen pathetici as their name might lead us to expect, and which only gives passage to small vessels). These nerves run direct to the superior oblique muscles to which they give motor power.

Abducens or sixth pair of nerves arise from the under surface of the medulla oblongata, just behind the pons Varolii, on either side of the inferior longitudinal fissure. Each nerve passes through foramen lacerum orbitale in company with the third, fourth, and ophthalmic division of the fifth nerves, and the ophthalmic artery of the side to which it belongs, and on gaining the orbit runs to supply the abducens muscle.

The **eyeball or globe of the eye** is a spherical body, slightly flattened from before backwards. The posterior three fourths is formed by a white fibrous coat, the sclerotic, its anterior part by a transparent layer, the **cornea**. To its postero-inferior and internal part runs the optic nerve, enclosed in its dense sheath of dura mater. The eyeball presents three coats, three cavities, and three humours, with accessory parts. The **sclerotic** is a fibrous coat, which is externally situated at the posterior three fourths of the eyeball. It is thickest posteriorly, and grows thinner to about the centre, when again it commences to grow thick, and anteriorly the cornea fits into it "like a watch-glass into its case. Around the opening of the optic nerve it becomes attached to its sheath, which becomes inflected inwards forming a plate, **lamina cribrosa**, through numerous perforations in which the fibres of the nerve pass. Through the largest of these perforations, termed the **porus opticus**, the arteria centralis retinæ passes. Numerous small foramina, through which the ciliary nerves and vessels pass, are to be seen on the surface of the sclerotic. Anteriorly it is directly continuous with the **cornea**, so that we cannot distinguish between them with the microscope, though the transparency of the cornea is sufficiently distinctive, and seems to be due to fluid contained between its laminæ, since opacity is caused by pressure. It is a segment of a smaller sphere than the sclerotic, which overlaps it most superiorly and inferiorly, so the part visible in the living animal is markedly ovoid in shape, the largest end being the inner. Around its outer circumferent margin we may observe a semi-opaque ring, *arcus senilis* caused by the presence of the *ciliary ligament* internally to this part. It consists of two parts, *cornea propria* and *cornea elastica*; the former is anteriorly placed, and its structure may be divided into several laminæ; the latter, posteriorly placed, consists of a single lamina. The *anterior surface* is covered by the transparent conjunctiva. Its *posterior surface* by the serous membrane of the aqueous chamber. The sclerotic must now be separated from the **choroid or middle coat** by laying the eyeball on a flat surface, and cutting through the most elevated portion of sclerotic with scissors, when the choroid, which was separated by the weight of its contents will be exposed. It is connected to the sclerotic by vessels and nerves which form **membrana fusca**, distinguishable as a few dark thread-like fibres running from one to the other. The choroid is a dark brown membrane, on the surface of which we see a number of light blue thread-like vessels ramifying, forming the external layer of the choroid (*venæ vorticosæ*, so called from a fancied resemblance to a whirlpool). The next layer mainly consists of arteries arranged in festoons, of which the convexities are anteriorly situated, *tunica Ruyschiana*. It rests upon a layer of polygonal pigment cells, which constitutes the *pigmentary layer*, and in which the rods and cones which form the external layer of the retina, are imbedded. Towards the anterior part of the sclerotic the choroid becomes reflected into the cavity of the eyeball, to form the "setting" for the crystalline lens; and, as the diameter of the lens is less than that of the chamber at this part, the choroid becomes puckered to accommodate it to the size of the parts. The puckerings are **ciliary processes**, the outer margin of which is termed the *ora serrata*. The inner margin comes in contact with the capsule of the lens to which it is attached, and between the two is situated a triangular canal, *the canal of Petit*. (Some uncertainty seems to exist as to the situation of this canal. I have followed Huxley.) The ciliary processes also contain some muscular fibres; anteriorly they are covered by the lining membrane of the aqueous chamber, posteriorly by the retina. They are continuous with the **uvea**, a layer of pigmentary matter which covers the posterior surface of the iris, and gives the eye its colour (deficiency of this constitutes "wall eyes"). The

Iris is a muscular diaphragm, which projects into the cavity of the eyeball, leaving centrally an oblong opening, the **pupillary opening or pupil**. It consists of two orders of unstriated muscular fibres:—*circular*, surrounding the pupil, and which cause it to contract; *radiating*, which extend from the outer circumferent margin to the pupil, which they serve to enlarge. These are posteriorly covered by the uvea, which overhangs the pupil above and below, forming dark pigmentary masses visible in the living subject, **corpora nigra**. They are largest at the upper margin of the pupil, and frequently are wanting below. The anterior and posterior surfaces of the iris are covered by serous membrane continuous around the margin of the pupil. The outer margin of the iris is attached to the **ciliary ligament**, a structure containing unstriated muscular fibre, situated at the outer circumferent margin of the posterior surface of the cornea, and serving to connect together the cornea, sclerotic, iris, ciliary processes, and choroid; in its substance is the *canal of Fontana*. The **optic nerve** runs from the optic decussation at the base of the brain, through foramen optici, after emerging from which it is pierced by arteria centralis retinæ, a branch of the ophthalmic. It runs through the orbit with this artery in its centre, and invested by a stout sheath, which extends from the dura mater to the sclerotic. After piercing the lamina cribrosa, its fibres radiate on the inner surface of the choroid, forming the **middle layer of the retina, or tunica nervosa**, which extends as far as the outer margin of the ciliary processes, ora serrata. On its external surface it has the **tunica Jacobi or bacillary layer**, which consists of nerve-cells (*rods, cones, and ganglion cells*) imbedded in the pigmentary layer of the choroid. On its internal surface are blood-vessels, which in the living subject may be seen by means of the ophthalmoscope (**tunica vasculosa**). At the opening of the optic nerve into the eye, *optic disc or blind spot*, visible on careful inspection of the living subject, nerve-cells are deficient, while there is a preponderance of nerve-fibres. In the human subject the *macula lutea or foramen of Soëmering* presents exactly opposite conditions, nerve-cells being here most numerous. We have failed to distinguish it in the horse. The different layers of the retina are connected together by areolar tissue, which is continued over the posterior surface of the ciliary processes, forming the *zona ciliaris* or *zone of Zinn*, which extends as far as the capsule of the lens. At the inner and upper part of the choroid we distinguish a part presenting a peculiar greenish metallic lustre, **tapetum lucidum**, which by reflecting instead of absorbing the rays of light, enables the animal to "*see in the dark.*"

The **chambers of the eye** are three in number. The *anterior chamber* is situated between the cornea and the iris. It communicates with the *posterior chamber*, which is between the iris, ciliary processes, and the capsule of the lens, through the pupillary opening (this communication seems to be closed in the fœtus by a vascular membrane, *membrana pupillaris*, which, however, is merely the vascular anterior part of the capsule of the lens to which the iris is attached). These two chambers are lined by the **membrane of Demeurs or Descemet**, which, therefore, covers the posterior surface of the cornea, invests both surfaces of the iris, and is continued over the anterior surface of the ciliary processes and the capsule of the lens. It secretes the *aqueous humour*, a fluid consisting of water and a small quantity of saline matter and albumen dissolved in it. It serves to afford free play to the iris, to preserve the due convexity of the cornea, and to (slightly) influence the direction of the rays of light.

The *dark chamber* is the largest of the three, anteriorly bounded by the capsule of the lens, in all other parts it is circumscribed by the retina. The membrane lining it, the **hyaloid membrane**, also sends trabeculæ inwards, by which the cavity is divided into numerous cells, in which we find a fluid analogous to the aqueous humour, but which in consequence of this arrangement does not flow freely when the dark chamber is punctured; this is the *vitreous humour*, through its centre in the fœtus, from behind forwards to the lens runs a branch of arteria centralis retinæ. Between the aqueous and vitreous humours is situated the **crystalline lens (or third humour of the eye)**. It is enclosed in a capsule, the outer circumferent margin of which is attached to the free extremity of the ciliary processes and of the zone of Zinn. Its anterior surface is covered by the membrane of Demeurs, its posterior by the hyaloid membrane, and the connective continuation of the retina is attached along its margin. The lens itself is biconvex, being most convex posteriorly. It is composed of numerous concentric transparent layers, which increase in density from without inwards. In the centre is a circular nodule, consisting of three triangular portions. The *laminæ* consist of fibres denticulated at their margins and decreasing in size towards each extremity. By apposition of their denticulæ they form firm continuous layers, of which the external is liquid, and serves to allow movement of the lens in its capsule; it is termed the **liquor Morgagni**, but its liquidity is supposed to be the result of post-mortem change.

The *ciliary arteries* and *arteria centralis retinæ* supply the eyeball with blood; the *ciliary nerves* endow it with ordinary sensation; the *optic nerve* with the special sense of *sight*.

The first trace of the embryo mammal is a longitudinal layer of cells, and from this both above and below an arch is formed. The *inferior or hæmal arch* is mainly composed of soft structures which are supported by a bony framework; it forms the face, thorax and abdomen. The *superior or neural arch* subsequently becomes a bony canal formed of many segments connected together by various structures, which posteriorly comes to a point, anteriorly four of its segments, or vertebræ form a large ovoid cavity with continuous walls, the **CRANIUM**. Twelve bones combine to form the cranium, four pairs and four single bones:—

Frontal, parietal, squamous and petrous temporal bones. (Pairs.)

Occipital, ethmoid, sphenoid, and os triquatrum. (Single.)

These bones are for the most part formed of two dense layers between which is cancellated structure. The central portion contains large veins, and is termed the *diplöe*, the compact laminæ are the *vitreous tables*. The *posterior part of the cranium* is formed by the occiput through the centre of which the foramen magnum passes. Inferiorly on either side of this is a foramen condyloideum. The *roof of the cavity* is formed posteriorly by the occiput which is united anteriorly to os triquatrum (centrally) and to the parietal bones on either side. The latter also help to form the *lateral walls of the cavity*. Anteriorly the roof is formed by the frontal bones from which a plate projects downwards to the upper extremity of the crista galli process of the ethmoid bone, which forms the centre of the anterior surface and on either side of which are found the **ethmoidal fossæ (or olfactory sinuses)** which are bounded by the *cribriform plate of the ethmoid*, while the body of the ethmoid forms the lower part of the anterior surface. In forming the lateral part of the walls of the cranium the parietal bones articulate with the wing of the sphenoid and with the squamous temporal, and the latter, in its turn, is posteriorly almost completely separated by the petrous temporal bone from the occiput. The *floor of the cranium* is formed posteriorly by the basilar process of the occiput, from which the body of the sphenoid runs forwards to the body of the ethmoid. From the lateral part of the body of the sphenoid the alae run to articulate anteriorly with the ethmoid and frontal bones, laterally with the parietal and squamous temporal; posteriorly each presents a rough margin with several notches which combines with the basilar process of the occiput and the temporal (petrous) bone to form the foramen lacerum basis cranii.

Foramina of the cranium. *Foramen magnum*, formed wholly by the occiput. Affords attachment to the anterior margin of the dura mater of the spinal cord and gives passage to the spinal cord, which here commences from the medulla oblongata, and its membranes. Also to venous sinuses, the meningeal and basilar arteries, and the major portion of the spinal accessory pair of nerves. *Foramen condyloideum*, formed by the occiput at the side of foramen magnum, in front of the condyle. It gives passage to the twelfth, lingual, or hypoglossal nerve.

Foramen lacerum basis cranii, situated at the side of the basilar process of the occiput, formed by the occipital, sphenoid, and petrous temporal bones. In the human subject by a bony septum is divided into two parts (in the horse by a ligament); through the posterior run the glossopharyngeal, pneumogastric, and spinal accessory nerves; through the anterior, the inferior maxillary division of the fifth, internal carotid artery, and meningeal branches, and a dilated vein running from the cavernous sinus to the jugular at its commencement.

Foramen auditorium internum. A circular opening at

the centre of the cranial surface of the petrous temporal bone. It gives passage to the seventh and eighth nerves, and runs to the internal ear though it gives off a portion which passes to the middle ear. On the upper surface of the body of the sphenoid, laterally are broad channels through which the superior maxillary division of the fifth runs, and thus gains the *foramen rotundum*, which, in proceeding to the bottom of the orbit, unites with the *sub-sphenoidal foramen*, which passes through the wing of os sphenoides, and which gives passage to the internal maxillary artery. Thus through the anterior part of the foramen rotundum, the superior maxillary division of the fifth nerve and the internal maxillary artery pass. Just above this and situated between the ethmoid and the sphenoid bones is *foramen lacerum orbitale*, through which the third, fourth, ophthalmic division of the fifth and sixth nerves, and the ophthalmic artery and vein run (the ophthalmic vein connects the superior varicose vein of the face with the cavernous sinus). In front of the sella turcica or upper surface of the body of the sphenoid, there are two foramina running forwards through the ethmoid bone connected anteriorly by a space which is covered by the inferior extremity of the cristagalli process and in which the optic decussation rests. These are the *foramina optici*. They give passage to the optic nerves covered by their dense sheath of dura mater, and open into the lower part of the orbits by a common opening, which they share with the foramina lacera orbitale, and rotunda. This is the **sphenoideal hiatus**, just above its superior rough margin is *foramen pathetici*, which only gives passage to a few small vessels. Internally situated is *foramen orbitale internum*, which passes to the outer lateral part of the ethmoidal fossa, thus passing directly into the cranial cavity. Through it the lateral nasal artery and vein run. Numerous other foramina are found passing through the cranial walls, but are unnamed; through them pass vessels and nerve-fibres. Some others are generally enumerated as cranial foramina, though they do not pass into the cranial cavity. *Foramen auditorium externum* runs through the external auditory process, and at the middle ear, in the recent subject, presents the membrana tympani. It is lined by common integument. *Foramen styloideum* is that opening through which the facial nerve emerges from the middle ear; it is divided into two parts; from the superior the nerve emerges, through the inferior the Eustachian tube communicates with the middle ear. It is just at the root of the styloid process of the petrous temporal bone. The inner surface of the os triquatrum presents an eminence (ossific tentorium) consisting of the convergence of three irregular bony ridges from which prominences run. On either side of the wall of the cavity one runs to the floor of the cranium, and one runs forwards along the central line to the crista galli process. In other parts the roof and sides of the cranium present slight prominences which separate broad shallow depressions (*digital impressions*) which are adapted to the convolutions of the brain.

The brain is covered by three membranes—*dura mater, arachnoid, and pia mater*. The **dura mater** is externally placed, and is a dense fibrous membrane, which forms the internal periosteum of the skull. It presents processes, some of which run outwards, others inwards. Those passing outwards establish its connection with the periosteum of the outer surface of the bones by passing through the immovable joints between them; this accounts for the difficulty experienced in separation of the dura mater from the bones at the sutures. The processes running inwards are two in number. **Falx cerebri** is a sickle-shaped fold, running from the superior extremity of the crista galli, and increasing in size until it terminates posteriorly in becoming attached to the ossific tentorium. It is broadest above and grows narrow below, forming a sharp edge. It fits into the superior longitudinal fissure of the brain. Along its attached margin runs the *superior longitudinal sinus*, along its free margin the *inferior longitudinal sinus*. These are connected together posteriorly by the *straight sinus*, into the anterior part of which *vena Galeni* empties itself, after running from the velum interpositum and from the upper surface of the corpus callosum. These sinuses are dilated veins situated in the dura mater, which splits to accommodate them. They are lined by epithelium continuous with that of the other veins, and at various parts we see fibrous cords (*cordæ Willissii*), which run across them to prevent undue distension. From the lateral parts of the ossific tentorium run folds to the floor of the cranium, the *tentorial membrane*. They are attached to the inner ridges of the parietal bones, and project into the great transverse fissure, in which they terminate in a sharp edge. In their substance run the *lateral sinuses* from the *torcular Herophili* above to the cavernous sinus. The fifth nerve (sensory portion) passes into them about their centre, and they thus contain it with its gasserian ganglion. The **torcular Herophili**, therefore, is a venous sinus, situated against the ossific tentorium, which receives the straight (and superior longitudinal) sinuses, and from which the lateral sinuses and also the *occipital sinuses* run. The latter pass backwards over the roof of the cerebellal cavity. Between the fifth nerves, at the base of the brain, the layers of the dura mater separate, forming the *cavernous sinus*, and imbedding a peculiar reddish-black body, circular, and flattened from above downwards, the use of which is unknown, but which receives the inferior extremity of the tuber cinereum—the **pituitary body**. Around the posterior and lateral parts of this body is situated the cavernous sinus lined by epithelium, to which run the lateral sinuses superiorly, the *anterior petrosal* anteriorly, and the *posterior petrosal* posteriorly, from the floor of the cranium; the latter blend with the sinuses of the floor of the spinal canal posteriorly. It also receives anteriorly the posterior extremity of the superior varicose vein of the face, which passes through foramen lacerum orbitale. Inferiorly it sends off prolongations, which pass through foramen lacerum basis cranii, and from which the jugular veins commence. In the cavernous sinus run the internal carotid arteries, covered by reflections of the epithelial lining layer; they pass from foramen lacerum basis cranii, communicating by a transverse branch behind the pituitary body, and emerging from the dura mater at the base of the brain, opposite the commencement of the fissure of Sylvius. The dura mater of the brain is continuous with that of the cord through the medium of its attachment to the margin of the foramen magnum of the occiput. It is attached to almost all the foramina around their margins, and is continued over the optic nerve as far as the sclerotic, where it terminates in forming the lamina cribrosa. On the outer surface of the dura mater, especially at the attached margin of the falx cerebri, have been distinguished minute nodular bodies, largest in old subjects, *glands of Pacchioni*. The **arachnoid** is a serous membrane, and presents a parietal and a visceral portion. The *parietal portion* lines the dura mater; the *visceral portion* invests the pia mater. Between these layers is a serous fluid, the *arachnoid fluid;* and between it and the pia mater is the *subarachnoid fluid.* Posteriorly this membrane forms the posterior part of the roof of the fourth ventricle, *arachnoidean valve;* it becomes reflected from the different prominences of the brain, and does not dip between into the sulci. Subarachnoid spaces are most marked in front of and behind the optic decussation and below the posterior lobes of the cerebrum in the transverse fissure. The **pia mater** is the vascular membrane of the brain, the surface of which it closely invests. It is prolonged into the cavities of the brain in several places; thus, in passing through the transverse fissure, it runs forwards under the posterior part of the cerebrum and becomes thicker, forming the *velum interpositum*, which is the roof of the third ventricle, and which anteriorly bifurcates, forming the *plexus choroides of the lateral ventricles*, which, passing through the foramina of Munro (anterior communicating foramina) and between the hippocampi and thalami optici, extend into the posterior cornua of their respective ventricles. Underneath the posterior part of the cerebellum it is similarly reflected into the fourth ventricle, forming the *plexus choroides of the fourth ventricle.*

The **brain** is primarily divided into *cerebrum, cerebellum, pons Varolii, and medulla oblongata.* The **cerebrum** is the largest part, and is that ovoid mass situated in the anterior part of the cranium. It presents on its external surface a number of *convolutions, gyri*, separated by *grooves, sulci*, which in the human subject are capable of division into sets, but in the horse present no regularity, except that at the middle line we may observe the "marginal convolution." On each side of the superior longitudinal fissure, and on opening the fissure, we may see a "gyrus fornicatus" on each side of the bottom of it. The cerebrum is separated from the cerebellum by the *great transverse fissure*, which extends on either side, and also in a forward direction, underneath the posterior lobes of the cerebrum, where it contains the vellum interpositum. The cerebrum is divided into two parts, *hemispheres*, by the *superior longitudinal fissure*, which runs forward from the transverse fissure and anteriorly extends downwards, putting in its appearance at the anterior part of the base of the brain;

and anteriorly and posteriorly presents somewhat firm union of the hemispheres. It contains the falx cerebri, and is terminated below by a layer of white matter, **corpus callosum**, which connects the two hemispheres, and is composed of fibres which run transversely from the white substance of the hemispheres, the anterior fibres being curved forwards, the posterior backwards, so that the narrowest portion is that shown by separation of the two hemispheres without incision. On its upper surface may be distinguished a central longitudinal line, the *raphé*, along which *arteria corporis callosi* passes, and on each are the lineæ longitudinales laterales, while lineæ transversæ intersect these at right angles. The corpus callosum terminates both anteriorly and posteriorly by a rounded extremity; the anterior is the genu, the posterior the splenu. From the latter the **fornix** runs in a forward direction, connecting together the hippocampi. Anteriorly, at the foramen of Munro, the fornix terminates in *four crura*. The *anterior* run to the base of the brain, where they terminate in the single corpus albicans (there are two in the human subject). The *posterior* take the name of corpora fimbriata, and extend into the posterior cornua of the lateral ventricle. The upper surface of the fornix is connected to the under surface of the corpus callosum by the *septum lucidum*, a thin layer of nerve matter which separates the lateral ventricles one from the other, and between the two layers of which is the *fifth ventricle* (which we have never found in the horse). By removing a small layer from the upper surface of the cerebral hemispheres we may demonstrate that they are composed externally of grey matter, internally of white. The grey matter, *cineritious*, is of uniform thickness throughout, but since it invests the convolutions presents a wavy margin. The *white or medullary matter* in this section is termed the *centrum ovale minor*, and presents the cut extremities of vessels, through which, in the fresh subject, a small quantity of blood oozes, *puncta vasculosa*. If sections are now made through both hemispheres on a level with the corpus callosum, the whole of the upper surface of that layer or *centrum ovale major* is exposed. By incision through this on either side we lay open the cavity of the **lateral ventricle**, a space divisible into *anterior and posterior cornua*, bounded anteriorly and externally by the substance of the hemispheres, superiorly by the corpus callosum, internally separated from its fellow by the septum lucidum, in the anterior part of which is the *foramen of Monro or anterior communicating foramen*. Its **floor** presents the following structures in order from before backwards:—

Corpus striatum, tænia semicircularis, plexus choroides, corpus fimbriatum, hippocampus.

Corpus striatum is a pear-shaped body, the base of which lies against its fellow, separated by brain substance, the apex extends into the anterior cornu, which runs downwards, outwards, and forwards. Externally it is covered by a thin layer of white matter, but is dark in colour, and has a few small vessels at its posterior margin. By passing a pin through it we shall find that it is situated just above locus perforatus anticus at the base of the brain. It is named from its composition of alternate layers of grey and white matter. At its posterior margin is distinguishable a thin layer of white matter, *tænia semicircularis*, which is generally covered by the plexus choroides which has been already described, and which also covers the posterior crus of the fornix, *corpus fimbriatum*, a thin layer which runs downwards into the posterior cornu, where it terminates in a small enlargement, pes accessorius. Immediately behind it is the *hippocampus*. (There are two, major and minor, in each lateral ventricle of man.) This is an internal convolution of the cerebral hemispheres, it rests upon the velum interpositum inferiorly, and, with the plexus choroides and corpus fimbriatum extends into the posterior cornu, where it terminates in the pes hippocampi.

The *posterior cornu* runs backwards, outwards, and downwards, and then forwards and inwards. It is shaped like a ram's horn, and its apex is situated in the mastoid lobule at the base of the brain. The hippocampus is separated by the velum interpositum from the *thalamus opticus*, a somewhat irregular, rounded body, which constitutes the deep seated origin of the optic nerve. It presents on its upper surface three prominences, *corpora geniculata externum and internum, and the tubercle*. The inner surface of this body is connected with its fellow by the *middle or grey commissure*, anteriorly situated to which is the foramen of Monro or anterior communicating foramen, which not only connects the cavities of the lateral ventricles but also extends down to the base of the brain, where it forms the cavity of the tuber cinereum, and the infundibulum, and extends as far as the pituitary body. Anteriorly it is bounded by the *anterior white commissure*, which connects the two corpora striata together. Posteriorly situated to the grey commissure is the *posterior communicating foramen*, which forms the commencement of the iter a tertio ad quartum ventriculum. Just above this is a small black cone-shaped body, the *pineal gland*, the use of which is unknown—we sometimes find it much enlarged. It contains pigmentary matter and vessels arranged in an areolar stroma, and from it forwards along the inner margins of the thalamic optici, above the grey commissure, run the *crura of the pineal body*, which anteriorly blend with the substance of the thalami. The **third ventricle** is the cavity through which the grey commissure passes "like a transverse beam." Anteriorly, it is bounded by the anterior white commissure, superiorly by the velum interpositum, laterally by the thalami optici and the crura of the pineal body. Inferiorly by the upper surface of certain structures situated at the base of the brain—tuber cinereum, corpus albicans, and locus perforatus posticus. It presents three openings: inferiorly, the infundibular; supero-anteriorly, the anterior communicating foramen; supero-posteriorly, the posterior communicating foramen. Its posterior boundary is the *posterior white commissure*, which connects together the *corpora quadrigemina*, and forms the roof of the *iter a tertio ad quartum ventriculum*. The **corpora quadrigemina** are four bodies posteriorly placed to the pineal body. The *anterior pair* is the roundest, largest, and darkest, and is termed the *nates*, the *posterior, or testes*, more ovoid in form, posteriorly receive tracts of white matter running from the cerebellum, *iter e cerebello ad testes*. These tracts are connected together by a thin membraniform layer of matter, valve of Vieussens, from which the fourth pair of nerves arise. They circumscribe anteriorly the **fourth ventricle**, "a boat-shaped cavity," situated at the upper part of the medulla oblongata. Its roof is formed anteriorly by the valve of Vieussens, centrally by the under surface of the cerebellum, posteriorly by the arachnoidean valve. Its floor and lateral walls are formed by the pillars of the medulla oblongata, anteriorly its cavity is continuous through the *aquaduct of Sylvius or* *iter a tertio ad quartum ventriculum* with the third ventricle. Its posterior part, from its likeness to the nib of a pen, is termed *calamus scriptorius*, and is continued through the spinal cord by a canal which extends centrally between its columns. The **cerebellum** is a round body, situated in a special division of the cranium, posteriorly to the cerebrum from which it is separated by the ossific and the membranous tentoria; and to which it is connected on each side by the iter e cerebello ad testes. It presents, on either side of the median line, running from behind forwards a fissure, and thus the organ is divided into three parts. Centrally is a "caterpillar-shaped" body, the anterior part of which is the *anterior vermiform process;* the posterior part, the *posterior vermiform process*, on either side of this is a **lobe**. By an incision from behind forward in the median plane we may expose the internal structure of the organ. The surface of the cerebellum presents convolutions which are divided into uniform parts by transverse secondary grooves. The grey matter which the external surface presents, is arranged with regard to the white as the leaves of a tree are to the branches, so the white matter inferiorly terminates in a broad base. This arrangement is the *arbor vitæ*, and the base of the stem is formed by three tracts of white matter, the **crura of the cerebellum**. The *anterior* is the iter e cerebello ad testis, the *posterior* is continuous with the corpus restiforme of the medulla oblongata, the *middle* passes under the medulla oblongata, joining with its fellow to form a transverse band at the base of the brain, the **pons Varolii**; this is the part which serves to connect together the three other divisions of the brain. It presents a longitudinal groove along its central line, through which the basilar artery passes, and on either side of which it becomes more prominent, for here is a collection of nerve-cells in its substance constituting the *corpus dentatum*, the special centre of the fifth pair of nerves which arises from the lateral parts of its under surface.

The **medulla oblongata** is but the commencement portion of the spinal cord, from which it is distinguished by its greater size, caused by the presence of a few extra nerve cells collected to form the *corpus olivare*. In addition to this, like the spinal cord, it presents three tracts of white matter (externally) continuous with those of the cord respectively. The *corpora pyramidalia inferior* are separated from each other by the inferior longitudinal fissure, which, however, is slightly obscured by the decussation or passage to the opposite side of the medulla of the motor fibres of the *corpora restiformia*, the lateral tracts of the medulla oblongata continuous with those of the spinal cord. These divide anteriorly, one branch runs to the cerebellum forming its posterior crus, the other passes forwards to the cerebrum and unites with fibres from the pyramids forming the crura cerebri. These restiform bodies are separated from the inferior pyramids by the *corpora olivaria*, small round bodies which internally present an irregular grey nodule, corpus dentatum. Their upper surface forms the floor of the fourth ventricle, and on either side superiorly is situated a small *corpus pyramidalium superior*. The two sides of the medulla oblongata are connected together by a commissure formed of grey and white matter.

The **base of the brain** presents numerous and diverse structures. At the anterior part centrally is the superior longitudinal fissure, and on either side of this is an **olfactory bulb**. This is a rounded mass of grey matter situated in the olfactory sinus, from which grey fibres run forwards through foramina, in the cribriform plates. This is attached to the under part of a tract of white matter which runs backwards and becomes narrower, terminating in two parts which separate at an angle enclosing a triangular space, *locus perforatus anticus, or substantia perforata*, the outer surface of the corpus striatum. These tracts of matter contain a cavity lined by serous membrane, which in some subjects is said to have been found continuous with the lateral ventricle. From the lateral parts of the thalamus opticus and from the corpora quadrigemina tracts of white matter run round the lateral parts of the crura cerebri. At the base of the brain they unite, and the fibres from this union form two nervous cords, the optic nerves which pass to the eye. This is the **optic decussation**, the fibres from one side pass—to the nerve of the opposite side, to the nerve of the same side, to the other optic tract; fibres also run from one optic nerve to the other. From the postero-lateral parts of the optic decussation the *crura cerebri* run to the anterior margin of the pons Varolii. They are broad bands of white matter, and from their under surface the *third pair of nerves, motores*

SPECIAL ANATOMY. 161

oculorum, arises. Between these, anteriorly is the *tuber cinereum*, which is continued to the pituitary body as the *infundibulum*, and through which a canal runs. Immediately behind this is the *corpus albicans*, formed by the anterior crura of the fornix, and more posteriorly, between the crura cerebri in the human subject, is *locus perforatus posticus, which in the horse is represented merely by a groove*. The hemispheres of the cerebrum of man are divided into anterior, posterior, and middle lobes. At the lateral part of the crura cerebri of the horse is a smooth, round mass, **mastoid lobule** (analogous to the middle lobe of man). From the optic decussation in front of this the *fissure of Sylvius* runs outwards, tending to divide the hemisphere into an anterior and a posterior lobe. Before running over the under surface of pons Varolii the **basilar artery** gives off the *anterior, posterior, and middle cerebellal arteries*, after doing so it terminates in receiving the *two posterior communicating arteries*. These give off the *posterior cerebral arteries* (which run upwards winding round the crura cerebri) and are given off with the anterior communicating and middle cerebral by the *internal carotid* after it has pierced the dura mater. The *anterior communicating artery* runs inwards and blends with its fellow in front of the optic decussation. The *middle cerebral* runs outwards through the fissure of Sylvius. From the anterior communicating or from the middle cerebral the *ophthalmic arteries* pass through the foramina lacera orbitale to the orbit, and from the centre of the anterior communicating the *anterior cerebral artery* (double in the human subject) passes around the anterior extremity of the corpus callosum into the superior longitudinal fissure, where it terminates in forming the *arteria corporis callosi*. Thus we have the **circle of Willis** at the base of the brain formed anteriorly by the anterior communicating, laterally and posteriorly by the posterior communicating arteries; it encloses the optic decussation, tuber cinereum, infundibulum, corpus albicans, locus perforatus, and part of the crura cerebri, and surrounds the pituitary body. We have always found the origin of the third pair of nerves externally placed to it. Posteriorly at the base of the brain we see the *pons Varolii* running transversely, and a *fifth nerve* arising at either end of it. From the under surface of the medulla oblongata, just behind the pons, on either side of the inferior longitudinal fissure, a *sixth or abducens nerve* arises; and more outwards and posteriorly placed the *glosso-pharyngeal, pneumogastric, spinal accessory, and lingual nerves*, while the *seventh and eighth* together arise immediately behind the fifth from the lateral part of the medulla. The **eleventh or spinal accessory** nerve arises from the lateral part of the spinal cord as low down as the fifth cervical vertebræ. It runs towards the cranium, passing between the sensory and motor roots of the spinal nerves, receiving fibres from the lateral part of the cord. Its last fibres arise from the medulla oblongata, it then emerges from the cranium through the posterior part of foramen lacerum basis cranii, and runs to supply sterno-maxillaris, levator humeri and other muscles of the neck. It has extensive communication with the nerves near its origin and with the superior cervical ganglion. The lingual nerve passes between it and the glosso-pharyngeal.

In laying open the cranium to examine its contents remove all surrounding soft structures. Saw through the forehead transversely about three quarters of an inch above the orbital process of the frontal bone. From the extremity of this incision divide the bones on either side so as to remove the superior third of the foramen magnum of the occiput. Raise the separated portion, *calvarium*, with the chisel, the dura mater with its falx cerebri will be seen attached to its internal surface. The anterior part of the spinal cord having been now cut through, the brain may be raised, and the nerves passing from its base divided in turn from behind forward. The olfactory bulbs will generally remain *in situ*. The contents of the cranium is termed the **encephalon.**

The **internal and middle ear** may be examined by immersing the petrous temporal bone in a solution of hydrochloric acid, three drachms of the acid to a quart of water. (We have never tested this process practically, but have seen it recommended.) After a time the bone will become deprived of earthy matter so that it can be divided by a common scalpel. The foramina should be first stopped with wax, that the ossicula auditûs may remain uninjured. The *petrous temporal bone* is the hardest bone in the body; it is, therefore, well calculated for communication of the vibrations of the atmosphere. In it is situated the organ of hearing, which, like all other organs of special sense, has certain nerve-cells to which the fibres of the nerve of that special sense run, and a special arrangement whereby those influences which cause the impressions are best able to alter the condition of the cells. The ear is divided into external, middle, and internal ears. The first has been already noticed; the **internal ear** is that part in which are situated the cells of hearing. It is divided into *cochlea, vestibule, and semicircular canals*. The **vestibule** is the central cavity. It has seven openings into it, those from the semicircular canals, one from the scala vestibuli of the cochlea, and one from the middle ear, fenestra ovalis. It contains a membranous sac, the *membranous vestibule*, which is divided into two parts. The *superior, utriculus*, receives the openings of the membranous semicircular canals. It is separated by a membranous partition from the *inferior or sacculus*, which has no openings. This membranous portion is in contact both externally and internally with a serous fluid. That surrounding it is the *perilymph*, that contained inside of it is the *endolymph* and contains numerous minute calcareous particles, *otoconites or otolithes*. The **membranous semicircular canals** exactly resemble the bony cavities in which they are placed. They contain endolymph and are surrounded by perilymph. They are three in number, *superior, internal, and posterior*, and each canal opens by its two extremities into the vestibule, one of these extremities is dilated to form the *ampulla*. The undilated extremities of the superior and internal canals open by a common orifice into the vestibule. To the ampullæ run branches of the eighth or auditory nerve.

The **cochlea** is a cavity hollowed out in the substance of the petrous temporal bone, shaped like a snail's shell, consisting of a canal arranged in a spinal manner around a central bony tube, *modiolus*. This canal is largest at its commencement, and superiorly comes to a point. It is divided into two parts by a spiral plate, which is internally attached to the modiolus and projects into the canal. The division is completed by a membrane consisting of two layers between which is the *scala media of the cochlea*. The two portions of the canal, termed respectively *superior or scala vestibuli, inferior or scala tympani*, communicate superiorly. The former inferiorly terminates in the vestibule, the latter at the fenestra rotunda, an opening in the bone closed by a double fold of membrane, which serves to bring about a communication between the cavities of the cochlea and of the tympanum. On the bony plate which separates the *scalæ* (*lamina spiralis*) the fibres of the auditory nerve are spread out; they terminate in cells which are situated in the scala media (*fibres or rods of Corti*), "arranged like the keyboard of a piano." On reaching the internal ear, after passing through foramen auditorium internum, the **auditory, portio mollis, or eighth pair of nerves** breaks up into three parts. One runs to the ampullæ, another to the membranous vestibule, the third passing into the modiolus gives off fibres outward into the lamina spiralis. The intensity of a sound is supposed to be distinguished in the vestibule, its "tone" in the scala media of the cochlea. The vestibule is connected with the cavity of the tympanum by the fenestra ovalis, which in the fresh subject is closed by mucous membrane and by the base of the stapes. It will be observed that the cochlea only of the bony labyrinth has its membranous counterpart attached to the bony walls, the membranous semicircular canals and vestibule being separated from the bone by perilymph.

The **middle ear or tympanic cavity** is situated between the external and internal ears. Its outer wall presents the *membrana tympani*, which closes the foramen auditorium externum. By the foramen styloideum it is connected with the pharynx through the medium of the Eustachian tube. It communicates inferiorly with certain cavities in the petrous temporal bone (similar to the sinuses of the face) the *mastoid cells*. Through this cavity the seventh nerve passes; it enters through foramen auditorium internum, and during its passage gives off the chorda tympani, which runs to the gustatory branch of the inferior maxillary division of the fifth, and also branches to the muscles of the middle ear; then it emerges by passing through foramen styloideum. The inner wall of the middle ear presents two foramina—*fenestra rotunda*, leading to the scala tympani; *fenestra ovalis*, leading to the vestibule. Into the latter fits the base of the **stapes**, one of a chain of bones which extends across the tympanic cavity to the membrana tympani. This bone is "stirrup shaped," and its arch superiorly is connected by fibro-cartilage with a portion of the *incus*. In this joint an osseous nodule may be observed, **os orbiculare**, *commonly enumerated as* "*the* *smallest bone in the body."* The **incus** resembles in shape a human molar, one of the fangs of which articulates with os orbiculare; the other becomes attached to the bony wall of the cavity. The crown has a synovial articulation with the upper portion of head of the **malleus**, the handle of which is attached to the inner surface of membrana tympani, vibrations of which are thus communicated to the bony chain, which, by the mechanical arrangement of its component ossicles, communicates them in an increased degree of intensity to the liquid contents of the vestibule, which produces impressions upon the auditory cells, made more powerful by the presence of the solid particles, otoconites, in the liquid. The bones of the middle ear are acted upon by muscles, of which we will only particularise the *stapedius*, running from the wall of the cavity to the stapes at its attachment to os orbiculare and *tensor tympani*, running from the margin of the bony ring surrounding the tympanum to the handle of the malleus at its attachment to the membrana. The middle ear is lined by transparent and thin mucous membrane, which covers all the structures contained in it, and is continued into the mastoid cells. It closes the foramen auditorium externum in forming the inner lining of membrana tympani, and also fenestræ rotunda and ovalis. It is continuous through the Eustachian tube with that lining the guttural pouch and the pharynx. The internal and middle ear are supplied with blood by the styloid division of the posterior auricular branch of the external carotid; they receive sympathetic fibres from the **otic ganglion**, a small collection of nerve-cells situated on the inner surface of the inferior maxillary division of the fifth nerve, just after it has emerged from the cranium through foramen lacerum basis cranii.

PART V.—SPECIAL ANATOMY.

The Neck.

FROM the incision along the central line of the front of the neck the operator dissects the skin from *panniculus carnosus* in an upward direction, it being moderately firmly attached until it arrives at the collection of celluloadipose material attached to the superior margin of the ligamentum nuchæ along the greater part of its length, which is somewhat intimately blended with the skin.

Extending from the root of the ear, parallel to the posterior margin of the lower jaw, is **parotideo-auricularis**, attached superiorly to the conchial cartilage, and running downwards to be attached to the surface of the parotid salivary gland. It is covered by a thin layer of panniculus. Behind the ear, along ligamentum nuchæ for about five inches, are attached the **retrahentes aurum** muscles drawing back the ear.

Along the lower part of the neck the muscular fibres of panniculus are distinct; anteriorly some of them assume a somewhat peculiar arrangement, running in a different direction to the other fibres, in a distinct band, towards the submaxillary space, meeting with the corresponding fibres from the opposite side; posteriorly the muscle becomes much increased in thickness, and sends some muscular fibres to cover the anterior part of the connection of pectoralis transversus with the cariniform cartilage; its main portion, however, extends over the shoulder with levator humeri, with the tendinous part of which muscle it blends, at the superior part of the neck, and so the panniculus carnosus becomes indirectly attached to the superior margin of ligamentum nuchæ. The splenius muscle may be seen after removal of the skin through the superior aponeurosis of the levator humeri.

On removal of panniculus we disclose, at the postero-superior part, the **sterno-maxillaris** shortly after its origin from the anterior extremity of the cariniform cartilage, blended with its fellow along the middle line, thus forming a sling underneath the trachea, about the anterior part of the posterior third of the neck, they separate, and, becoming round muscular cords, run forwards to be inserted each by a broad tendon, which, after separating the parotid from the submaxillary salivary glands, passes over stylo-maxillaris, and becomes attached to posterior margin of the lower jaw about three inches below its articulation with the squamous temporal bone.

The trachea at the middle third of the neck is covered by the **sterno-thyro-hyoideus** muscle, which, passing from the anterior extremity of the cariniform cartilage, here divides into two parts; the thinner runs underneath sub-scapulo-hyoideus to the lateral part of the external surface of the ala of the thyroid cartilage, where it terminates, being inserted between the muscles proper to the larynx: just above this termination is situated the **thyroid body**, a small dark-red body of a rounded form attached to the lateral aspect of the anterior part of the trachea. The thyroid body under the microscope presents a number of vesicular cavities lined by tesselated epithelium embedded in a somewhat soft stroma. It is one of the so-called vascular glands, the use of which is not distinctly understood, and receives a *large branch from the carotid artery*, the only constant one given off from that vessel in its course up the neck; this branch frequently sends off a twig to the larynx which receives the name of the *laryngeal artery*. The broader division of sterno-thyro-hyoideus runs towards the centre of the submaxillary space, where it becomes inserted into the under surface of the body of os sphenoides, together with **subscapulo-hyoideus**, which, however, occupies a much larger surface of attachment, and which puts in its appearance about the anterior third of the neck, running from underneath sterno-maxillaris to meet its fellow at its attachment. By removal of parotideo-auricularis and panniculus, **the parotid gland** is exposed. It is an irregular-shaped organ of a dirty white colour, composed of numerous minute lobules, which fit in between the different structures it covers. Anteriorly it extends a little on to the surface of masseter externus, being superiorly in contact with the temporal artery and vein and facial nerve as they arise and wind round the jaw, and then extends upwards over the muscles in front of the ear forming one of its superior divisions, the other being situated behind the ear. From this the posterior margin of the gland runs in a downward direction, being separated from the wing of the atlas by tendinous structure, and terminates at the junction of the submaxillary and jugular veins, from which point its lower margin runs forwards to the angle of the jaw. From this antero-inferior point the **duct** of the gland arises, and passes over the internal surface of the lower edge of masseter internus, where it is covered only by panniculus and skin, and is in company with the submaxillary artery, and the corresponding vein. Through the substance of the parotid run several large veins, among others the jugular, and by raising it and removing the **stylo-maxillaris**, running from the anterior part of the styloid process of the occiput to the angle of the lower jaw, blending superiorily with digastricus, we expose the termination of the

Carotid artery. Just opposite to the commencement of the trachea it breaks up into three parts, one,

Ramus anastomoticus, runs directly upwards to the under surface of the wing of the atlas (giving off a large branch running to the occiput), here it rests upon obliquus capitis anticus, and anatomoses with a vessel, the vertebral artery, coming through the posterior foramen in the wing of the atlas; the trunk so formed passes into the anterior foramen, which has two outlets, through one of which a branch runs to obliquus capitis inferior upon the upper surface of the atlas, through the other a branch passes into the spinal canal, and after piercing the dura mater of the spinal cord anastomoses with its fellow from the opposite side and runs forwards, as the basilar artery, to supply the brain. The smallest of the three terminal branches of the carotid, the **internal carotid**, runs directly towards the base of the cranium throughout its course being situated with its fellow at the posterior part of the septum formed by the contact of the internal walls of the two guttural pouches.

The **external carotid** seems to be the direct continuation of the carotid trunk, it passes forwards, and just before arriving at the posterior angle of the long cornu of os hyoides gives off the *submaxillary artery*, which, running between the long cornu and hyoideus magnus, gives off the *lingual artery* running forwards, and then meets with and accompanies the corresponding vein, and the parotid duct over the internal surface of the inferior margin of masseter internus. After reaching the external part of the dilated superior extremity of the long cornu, the external carotid gives off *large posterior masseter branches* anteriorly, *parotid branches* and branches to stylo-maxillaris externally, and the **posterior auricular** posteriorly. This latter branch gives off the **middle auricular**, which runs to the external auditory opening. The external carotid is here covered by the mucous membrane of the guttural pouch, a transparent layer which covers *both* sides of the long cornu of os hyoides; on reaching the temporomaxillary joint *the artery terminates in three branches*, the **anterior auricular**, which runs to the front of the ear to supply the structures in that neighbourhood; the **temporal**, which winds round the jaw and runs along the superior margin of masseter externus; and the **internal maxillary**, which runs to the base of the cranium.

The **levator humeri** must now be dissected to its attachment to the crest of the occiput; to the mastoid ridge extending downwards laterally from that ridge, the wing of the atlas by a tendon common to it and trachelo-mastoideus and splenius, the transverse processes of the second, third, and sometimes fourth cervical vertebræ. By turning up its inferior margin we show the branches of the *cervical intervertebral nerves* running directly to it. Its attachment to the superior margin of the ligamentum nuchæ by faschia, and its blending posteriorly with trapezius may also be noticed; by this time its attachments to the fore limb will be shown, *i. e.* to the spine of the scapula, the point of the shoulder, the humeral ridge with its elongating ligament, and the antero-inferior depression of the humerus.

By removal of levator humeri we expose a *second layer of superior cervical muscles*. Running from the transverse processes of the cervical vertebræ, as far forwards as the third to the superior margin of the scapula is *serratus magnus*, and from the antero-superior angle of the scapula *rhomboideus* runs in the form of an extremely elongated triangle to within six inches of the attachment of the cordiform portion of the ligamentum nuchæ to the occiput.

Seen between these muscles and covered at its posterior part by them is

Splenius, a large layer of muscle attached anteriorly, in common with trachelo-mastoideus, to the mastoid ridge, the crest of the occiput and the wing of the atlas, it also has heads attached to the transverse processes of the second, third, fourth and fifth cervical vertebræ. Posteriorly it is attached to the superior spinous processes of the three anterior dorsal vertebræ, and superiorly to the cordiform portion of ligamentum nuchæ along its whole length. It posteriorly sends a tendon to blend with the commencement portion of the dorsal faschia. Having removed this muscle we expose another muscular layer consisting mainly of complexus major, covered inferiorly against the vertebræ by

Trachelo-mastoideus, which is composed of three distinct portions: 1st. Head is attached with splenius to the mastoid process, and with complexus major to the oblique processes of the fourth, fifth, sixth and seventh cervical vertebræ, and the transverse processes of the two anterior dorsal vertebræ: 2nd. Head with splenius to the transverse processes of the two anterior and separately the oblique processes of the second, third, fourth, fifth, sixth and seventh cervical vertebræ; 3rd. Head with splenius to fourth cervical transverse process; the second and third heads join with the first to become attached posteriorly. Between the superior tendons the oblique muscles of the head are to be seen.

Obliquus capitis superior arises from the mastoid ridge and from the styloid process of the occiput, its fibres run obliquely downwards and backwards, being attached internally to the altoido-occipital capsular ligament, and become inserted into the anterior margin of the transverse process of the atlas, at the inferior margin of this muscle, slightly concealed by it, is a small muscular band,

Obliquus capitis anticus, which runs from the styloid process of the occiput to the under surface of the body of the atlas at its lateral part—it is also attached to the capsular ligament of the joint.

Obliquus capitis inferior passes from the supero-external part of the wing of the atlas, over the altoido-dentatal joint to the capsule of which it is attached, and becomes inserted into the lateral part of the superior spinous process of the dentata; posteriorly it is in contact with the anterior semi-spinalis muscle of the neck, and on the wing of the atlas it covers the vertebral artery, where it passes on to the ala, and then dips down through its posterior foramen to anastomose with ramus anastomoticus. Through the anterior foramen a branch of the artery runs from the union of these two to the muscles situated about here,. but the main portion passes through the foramen in the pedicle of the atlas, and after gaining the vertebral canal, pierces the dura mater, and uniting with its fellow forms the basilar artery, which runs to the brain.

Complexus major is attached in common with trachelo-mastoideus to the oblique processes of the fourth, fifth, sixth and seventh cervical vertebræ, and the transverse processes of the two anterior dorsal vertebræ. It is also attached to the oblique processes of the third, fourth, fifth and sixth dorsal vertebræ, and to the oblique process of the third cervical vertebræ; anteriorly it blends with complexus minor, and becomes inserted into the scabrous pit of the occiput. It is attached along the greater part of the superior margin of ligamentum nuchæ. Its posterior attachment is covered by a muscle which seems at first sight to be a part of longissimus dorsi, *i.e.* **cervico-dorsalis**, which is attached in common with complexus major to the oblique processes of the six anterior dorsal vertebræ, and runs forwards to the transverse processes of the three posterior cervical vertebræ (to the seventh it is attached by a tendon, which it shares with longissimus dorsi).

Complexus minor arises from the posterior part of the superior spinous processes of the dentata, and, blending anteriorly with the complexus major, becomes inserted into the scabrous pit of the occiput. It lies in contact with its fellow of the opposite side, for about here the ligamentum nuchæ is deficient, there being a space between its cordiform and membraniform portions. By removal of it we expose

Rectus capitis posticus major, running from the anterior part of the superior spinous process of the dentata, becoming attached to the atloido-axoidal capsular ligament, and blending in, become inserted into the scabrous pit of the occiput with

Rectus capitis posticus minor, which arises from the superior part of the arch of the atlas, and is attached to the atloido-occipital capsular ligament.

The **jugular vein**, after passing through the parotid gland, receives the submaxillary vein when about opposite the thyroid body. It then proceeds down the channel of the neck, receiving no important branches, until it arrives at the lower part, when the *superficial brachial vein*, which carries blood from the fore extremity opens into it, and in some cases it receives the lymph from the *thoracic duct* on the left side, the great right lymphatic on the right (often, however, this opens into the axillary vein). Finally, between the two first ribs the jugulars unite to form the commencement portion of the anterior vena cava. By removing the jugular vein and subscapulo-hyoideus we expose the **trachea**, and on its supero-lateral part the carotid artery with the pneumogastric and gangliated cord of the sympathetic nerves enclosed in its sheath, and the inferior laryngeal branch of the pneumogastric a little lower down.

The **Œsophagus** is a long tube which commences anteriorly at the posterior part of the pharynx, and after passing along the neck, and through the thorax, and the foramen sinistrum of the diaphragm, terminates at the cardiac or left opening of the stomach. At first situated above the larynx in its course down the neck, it becomes inclined to the left side (thus we watch the left side of the neck to see a ball " go ") so that on entering the thorax it is near the inner surface of the first left rib. While passing through the anterior mediastinum it becomes slightly above the trachea inclined to the right side, and in passing over the base of the heart it crosses the right bronchus, then in the posterior mediastium, having rested in a special groove in the left lung, where it is in company with the pneumogastrics, it passes with them through foramen sinistrum. It consists of two coats. The external is muscular, anteriorly it is directly continuous with crico-pharyngeus; it consists of longitudinal fibres externally placed, circular fibres internally, and until it has passed over the base of the heart it retains its red colour, which is apt to cause it to be mistaken for some of the cervical muscles in the neck. Posteriorly it becomes thicker, and its fibres are continuous with those of the stomach. This coat is connected to the *internal or mucous coat* by very loose areolar tissue, in consequence of which the latter is generally arranged in longitudinal folds which close the cavity of the tube. It is the termination of this mucous coat at the stomach which *is said to prevent vomition in the horse except* under extreme circumstances; others attribute it to the mode of entrance of the œsophagus into the stomach, while some argue the presence of a special spiral valve. This mucous membrane is similar in structure to that of the pharynx and to the cuticular portion of the mucous lining of the stomach, with which it is continuous anteriorly and posteriorly respectively.

In the upper and anterior part of the neck, perceptible on manipulation in the living animal, is the **larynx**, a remarkable cartilaginous organ the seat of the voice; the upper part of the trachea specially modified. It is composed of several pieces of cartilage, which are united together in some places by synovial, in others by fibrous articulations. Elastic expanded ligaments in some situations strengthen the connection. The larger cartilages are five in number; thyroid, cricoid, two arytenoids, and epiglottis. The **Thyroid** is the largest, and consists of a body and two wings. The *body* is supero-anteriorly placed, and is the thickest portion of the cartilage. It has two surfaces, *anterior* and *posterior*, and four margins, superior, inferior, and two lateral. To the internal surface the epiglottis is attached by elastic tissue, the external surface is in contact with the sterno-thyro-hyoidei muscles. To the superior margin is attached the *thyro-hyoid ligament*, to the inferior margin the *crico-thyroid ligament*, and the *alæ* or *wings* arise from the two lateral margins. The *alæ* are firm flat portions of the cartilage which exhibit a tendency to undergo ossification; each of them manifests the form of an oblique-angled parallellogram, the anterior acute angle of which blends with the body of the cartilage; the postero-inferior acute angle is rather elongated, and presents a synovial surface for articulation with the antero-inferior synovial facet of the cricoid cartilage. The postero-superior oblique angle presents a flexible cartilaginous prolongation, *superior cornu* of the posterior margin, the free extremity of which is united by fibro-cartilaginous tissue to the posterior extremity of the heel process of os hyoides. The antero-inferior oblique angle forms the point of junction of the anterior and inferior margins, to both of which is attached the crico-thyroid ligament. The superior margins of the two alæ are connected together by the superior extremity of the body and from the whole ridge thus formed the *thyro-hyoid ligament* runs to the posterior crescentic margin of the hyoid bone. This ligament fills the space between the superior cornu of the cartilage and the alæ, with the exception of the angle, where it is deficient, and thus a *foramen is formed through which the terminal portion of the superior laryngeal branch of the pneumogastric runs* to supply the mucous membrane of the larynx with exquisite sensibility. The external surface of the thyroid alæ is in contact with the thyro-hyoideus superiorly; thyro-pharyngeus supero-posteriorly; crico-thyroideus inferiorly; between these the short portion of sterno-thyro-hyoideus is attached. From the superior cornu the posterior part of hyo-thyro-pharyngeus runs upwards. The internal surface is superiorly covered by mucous membrane, inferiorly by thyro-arytenoidei posticus and anticus (and is in contact with crico-arytenoideus lateralis). The **Epiglottis** has aptly been likened to a sage leaf, the base of which is attached by elastic tissue to the internal surface of the body of the thyroid, the apex is invested closely by mucous membrane, and is free. It consists of yellow elastic cartilage, and the mucous membrane covering it presents very large mucous glands. To the anterior surface mucous membrane proceeds from the base of the tongue, and by removal of this centrally we shall expose

Hyo-epiglottideus running from the posterior part of the body of os hyoides, directly backwards to the anterior surface of the epiglottis. The base of the epiglottis on either side is continued by a prolongation which runs in a backward direction, connected by cartilaginous nodules (called the cuneiform cartilages) to the arytenoid cartilages. The posterior surface of the epiglottis is covered by mucous membrane, and during the act of deglutition is brought down over the glottal opening or entrance to the larynx.

The **cricoid cartilage**, as its name indicates, is extremely like a finger-ring. Its smallest part is anteriorly situated, and from this it increases in size irregularly in a posterior direction. The posterior part forms a flat hexagonal surface, with a central ridge and four synovial articulatory facets, two at the antero-inferior angles for articulation with the posterior acute angles of the thyroid alæ, two at the postero-superior angles for articulation with the arytenoid cartilages. On either side of the central ridge is attached a crico-arytenoideus posticus, while crico-arytenoideus lateralis originates from the posterior part of the superior margin of the smaller portion of the cartilage. The whole inferior margin affords attachment to the crico-trachealis ligament, the anterior part of the superior margin to the crico-thyroid ligament. The lateral parts of the external surface to the crico-thyroid muscles. The internal surface is lined by mucous membrane.

The **arytenoid cartilages**, two in number, articulate inferiorly by synovial joints with the cricoid cartilage. The articulatory part is shaped like a prism, the inner angle of which is elongated in a posterior direction towards its fellow; the outer angle in an anterior direction towards the epiglottis, with which it is indirectly connected by the above-mentioned cartilaginous nodules. The superior extremity of the prismatic portion is expanded, and sends a peculiar thin process in an anterior direction to be united by mucous membrane to the epiglottis. Thus the external surface presents two triangular spaces, into the posterior of which arytenoideus is attached (running from one cartilage to the other), into the anterior the thyro-arytenoidei posticus and anticus, while crico-arytenoidei lateralis and posticus are attached to the prominent ridge between the two. The superior margins of the arytenoid cartilages are connected by continuity of their investing mucous membrane.

From the internal surface of the arytenoid cartilages to the inner surface of the thyroid cartilage and of the crico-thyroid ligament run bands of yellow elastic tissue termed the **vocal cords**, by the tension of which the sound of the voice is modulated. They are covered on their internal surface by the laryngeal mucous membrane, and externally are in contact with the thyro-arytenoidei muscles; the triangular space between them is the rima glottidis. In our description of the cartilages we have incidentally noticed the different layers of yellow elastic tissue which, while they serve to connect the different portions of cartilage together, also by their elasticity cause them to return to their normal position after allowing them to accommodate themselves to the movements of the neighbouring parts. The crico-thyroid, crico-trachealis, crico-arytenoid, and thyro-hyoid ligaments therefore require no further description. The mucous membrane of the postero-inferior part of the pharynx, after being reflected over the margin of the arytenoid and epiglottic cartilages before investing the vocal cords on either side, presents a diverticulum, the ventricles of the larynx, which are supposed to give resonance to the voice; they pass down between the thyro-arytenoidei anticus and posticus, and in the upper part of each in the ass we see a small opening leading into a secondary cavity. After passing over the cordæ vocales the mucous membrane lines the inner surface of the cricoid cartilage, and is continued downwards into the trachea.

Thyro-hyoideus arises from the external surface of the wing of the thyroid, and runs to be attached to the heel process of os hyoides; its internal surface is in contact with the thyro-hyoid ligament.

Crico-thyroideus runs from the external surface of the lateral part of the cricoid to the posterior part of the inferior border of the wing of the thyroid.

Crico-arytenoideus posticus arises from the superior surface of the posterior hexagonal part of the cricoid, and its fibres converge to become inserted into the central ridge on the external surface of the arytenoid, in common with

Crico-arytenoideus lateralis, which arises from the superior margin of the cricoid at the lateral part. **Arytenoideus** runs from the external surface of one arytenoid to that of the other, consisting really of two muscles blending along the central line.

Thyro-arytenoidei posticus and anticus (generally considered as one muscle) are separated by the inflection of mucous membrane, which forms the ventricle of the larynx. They arise from the internal surface of the thyroid, and run to the anterior part of the external surface of the arytenoid.

The **trachea** is a large tube through which air passes in its passage from the larynx into the lungs and *vice versâ*. It is composed of a series of cartilaginous rings, fifty to sixty in number, which are embedded in and connected together by a quantity of yellow elastic tissue. These rings are thickest inferiorly, and superiorly are deficient, being terminated by two thin expanded extremities which overlap each other, and are connected together by loose areolar tissue. Anteriorly the trachea commences by being connected to the inferior margin of the cricoid cartilage by the *crico-trachealis ligament*, which allows considerable motion between the two. Posteriorly after crossing the base of the heart in the thorax, where it is situated between the large arteries at their commencement and the large veins at their termination, it terminates, forming the two bronchi, round cartilago-elastic tubes running to the lungs. Its last cartilaginous ring is different from the rest, since it forms a species of sac from either side of which a bronchus passes, the right opening being the largest. The trachea therefore passes down the neck slightly inclined to the right side; superiorly it is in contact with the longi colli muscles, laterally with the carotid arteries, the pneumogastric, recurrent, and sympathetic nerves, and on the left side with the œsophagus. The jugular veins are in contact with its lateral part superiorly and inferiorly, but at the centre are separated by the subscapulo-hyoidei muscles. Sterno-maxillaris and sterno-thyro-hyoideus respectively, by blending with their fellows, help to support it. In passing between the first ribs it is superiorly placed to the commencement portion of the anterior vena cava, and to the termination of the arteriæ innominatæ, and here the œsophagus passes over it to gain the right side. At its extreme upper part the œsophagus is superiorly placed, and the thyroid bodies on either side generally connected by a central portion which runs round its inferior face. On the inner surface of the trachea, running from one side to the other at its superior part, is a continuous layer of transverse unstriated muscular fibres termed **trachealis transversus**; the outer surface of this muscle is connected to the cartilages by extremely loose areolar tissue, while its inner surface is covered by the mucous membrane which lines the trachea, and which presents ciliated columnar epithelium. The **pneumogastric nerve**, after emerging from the cranium through foramen lacerum basis cranii, and sending numerous fibres to the neighbouring nerves and ganglia, gives off the *superior laryngeal nerve;* it then runs down the neck in a sheath with the carotid artery and the recurrent, and sympathetic nerves. The *recurrent is the inferior laryngeal nerve, and supplies all the intrinsic muscles of the larynx with motor power except crico-thyroideus, which receives a branch from the superior laryngeal.* The left recurrent nerve is involved in the disease termed "roaring," whence paralysis and fatty degeneration of the muscles supplied by it ensues especially of the crico- and thyro-arytenoidei muscles. The **superior-laryngeal nerve** is compound at its origin; on arriving at the larynx it supplies some of the pharyngeal muscles, and thyro-hoideus and crico-thyroideus, and then becoming purely sensory pierces the foramen beween the thyroid cornu and the wing of the thyroid, and supplies the lining mucous membrane. The **carotid artery** in passing up the neck along the lateral part of the trachea gives branches to the œsophagus and other neighbouring structures, but the only constant branch is *one which runs to the thyroid body*, and which also sends off a large branch, *the laryngeal*, to the larynx. Along the under surface of the vertebræ runs **Longus colli**, which anteriorly extends to the tubercle or inferior spinous process of the atlas, and runs as far back as the sixth dorsal vertebra, becoming attached to the under surface of the bodies and transverse processes, and inferior spinous processes (except sixth cervical) of all the intervening vertebræ. It lies in contact with its fellow internally, and its thoracic portion is covered by pleura and forms a somewhat remarkable prominent cylindrical muscular mass. On its external surface posteriorly just in front of the first rib we see the vertebral artery and vein in their course between the canal through the sixth cervical transverse process and the entrance to the thorax. With them runs a nerve formed by union of a branch from each cervical nerve, through the middle, to the inferior cervical ganglion —both of which ganglia are situated between the two first ribs. These structures are hid from view by

Scalenius, a muscle composed of two distinct heads, the superior or shorter arises from the transverse process of the seventh cervical vertebra and runs to the anterior margin of the inferior extremity of the superior third of the first rib; the longer or inferior arises from the transverse processes of the fourth, fifth, and sixth cervical vertebræ, and its fibres converge to become inserted into the superior part of the inferior third of the external surface of the first rib. *Between the two heads the axillary plexus of nerves passes, below the inferior head, the axillary artery and vein, the latter being inferiorly situated.* On removal of this muscle we see the origin of the **axillary plexus** from the sixth, seventh, and eighth cervical, and first dorsal, and sympathetic nerves, and the **phrenic**, running in a direction backwards, having originated by fibres from the fourth, fifth, and sixth cervical. At the anterior part of the neck longus colli is covered by the

Rectus capitis anticus major, which arises from the transverse processes of the third, fourth, and fifth cervical vertebræ, and becomes inserted into the under part of the posterior extremity of the body of the sphenoid bone. Just behind this

Rectus capitis anticus minor is inserted into the antero-inferior part of the basilar process of the occiput, running from the under surface of the body of the atlas, and becoming attached to the atloido-occipital capsular ligament in its course. The inferior surface of these muscles is covered by the mucous membrane of the guttural pouch, and is in contact with numerous large nerves, the internal carotid artery, and the **superior cervical ganglion,** a grey nodular body, from which the gangliated cord of the sympathetic commences its cervical portion.

The supero-posterior part of the neck is supplied with blood by the **posterior cervical artery.** It commences from the antero-superior part of the arteria innominata, on the left side separately, on the right side by a root common also to the anterior dorsal artery. It passes upwards through the first intercostal space, giving off the *first intercostal artery.* It then passes forwards and upwards over cervico-dorsalis, and subsequently runs between ligamentum nuchæ and complexus major to anastomose with the artery of the poll.

The infero-posterior part is supplied by the **inferior cervical artery,** which shortly after its origin from the arteria innominata passes forwards, and breaks up in supplying the muscles in this neighbourhood — panniculus carnosus, pectorales transversus and anticus, sterno-maxillaris, &c. Three sets of intrinsic muscles of the neck require to be noticed:

Intertransversalis colli passes from the transverse process of one vertebra to that of the vertebra next in order to it.

Spinalis colli runs from the oblique process of one bone to the transverse process of the vertebra second in order behind it.

Semispinalis colli from the superior spinous process of one vertebra, passing over two bones, gains the oblique process of the third in order.

We have now arrived at the cervical portion of the spinal column; by sawing through the laminæ of the vertebræ on either side we may expose the **spinal cord**, *in situ*, covered by its membranes. The **dura mater of the cord** differs from that of the brain in that it does not form the periosteum lining the bony cavity in which it is placed. It is continuous anteriorly with the corresponding membrane of the brain, in passing to the margin of foramen magnum of the occiput. Posteriorly it forms the *filum terminale* in becoming attached to the posterior extremity of the bony canal. On either side it is closely attached to the intervertebral foramina, and is reflected across these openings to close them, so that here it is thick, and presents two rows of foramina, through which the collections of nerve-fibres pass. The **pia mater** is directly continuous with that of the brain and closely invests the cord. It in structure closely resembles that of the brain, and laterally is connected to the dura mater by the **ligamentum denticulatum**. This is a band of white fibrous tissue extending along the lateral part of the cord, attached to the external surface of the pia mater, between the superior and inferior roots of the nerves. It sends processes outwards opposite the bodies of the vertebræ, which become attached to the inner surface of the dura mater. The arachnoid membrane is continuous with the cranial arachnoid, lines the dura mater, covers the pia mater loosely so as to produce a *subarachnoid space*, and invests the processes of ligamentum denticulatum, and those portions of the nerves which pass through the arachnoid space.

The **Spinal cord** consists of a number of nerve-centres united together by white fibrous nerve cords, which render it a continuous and extremely elongated mass extending from the brain anteriorly, through the spinal column *as far as about opposite the second sacral bone*. It varies in size in different parts, and this variation is due to the number of nerves it must give off. Thus those parts which supply nerves to the fore and hind extremities are very large. It is moderate in size in the central part of the cervical region, but in the posterior cervical and anterior dorsal vertebræ it grows larger, and from this decreases in size until the lumbar region, where it becomes large in giving off nerves to the hind limb. As a rule the nerves pass straight outwards to the intervertebral foramina, but in the sacral region they run obliquely backwards, and the superior fibres unite with the inferior and again separate, forming the superior branches which pass through the supersacral, the inferior through the subsacral foramina. In all other regions this union and subsequent separation occurs after the nerves have pierced the dura mater in passing through the intervertebral foramina. This arrangement of the nerve-fibres in an oblique direction in the sacral region forms the *cauda equina*. The spinal cord consists of *two hemispheres* placed side by side, separated superiorly by the *superior longitudinal fissure*, inferiorly by the *inferior longitudinal fissure*, the latter is shallower and broader than the former. They are separated by a transverse band connecting the hemispheres, termed the *commissure*, which consists of grey matter with white substance on each surface, of which only the superior layer is thick enough to be visible to the naked eye. A transverse section of the cord shows that centrally it presents grey matter arranged like inverted commas in the hemispheres, connected by the grey commissure the convex surface of one to that of the other. Thus the *grey matter presents two cornua in each hemisphere*. The superior is the longest and narrowest, and runs upwards and outwards, reaching the surface of the cord. The inferior is rounded and separated from the surface by a layer of white matter. From the supero-lateral part of the cord to which the superior cornu approaches the superior nerve-fibres arise from the whole length of the cord, while from the infero-lateral part nearest the inferior cornu the inferior fibres arise. *The superior or sensory fibres* are divided into sets which converge towards the intervertebral foramina. They pass through the superior openings in the dura mater, have a ganglion in connection with them, and are microscopically indistinguishable from the *inferior or motor fibres*, which run from the cord through the inferior row of openings in the dura mater to subsequently unite with the superior fibres forming the spinal nerves. By this arrangement of the nerve-fibres each hemisphere of the cord is divided into a *superior, a middle and an inferior column*. The spinal nerves at the commencement send fibres or receive fibres from the nearest sympathetic ganglion. In most regions the ganglia of the sympathetic cord are close against the intervertebral foramina, but in the neck the fibres from each nerve unite to form a branch running with the vertebral artery through the foramina in the transverse processes of the vertebræ to terminate at the inferior cervical ganglion. The **cervical nerves** are eight in number. The **first or suboccipital** emerges through the foramen in the pedicle of the atlas, through which the vertebral artery gains the spinal canal. It sends fibres to the auricular nerves and to the superior cervical ganglion (situated underneath the ala of the atlas). The *second nerve* emerges through the foramen in the anterior part of the pedicle of the dentata. The *third, fourth, and fifth nerves* send fibres to form the phrenic nerve, while fibres from the *sixth, seventh, and eighth, first dorsal, and sympathetic* form the axillary plexus. Each of these nerves after passing through its foramen sends off a superior branch to the structures above the vertebræ, an inferior to those below. On the lateral part of the cord as far down as the fifth cervical vertebra may be seen the **spinal accessory** nerve which passes upwards, between the superior and the inferior roots of the nerves, receiving fibres along its whole course from the cord as far up as the medulla oblongata.

After removing the cord and its membranes we may examine the **superior vertebral ligament,** which extends along the upper surfaces of the vertebræ throughout the whole canal. It becomes widened in becoming attached to each intervertebral cartilage and is attached to the roughened line on the centres of the upper part of the bodies of the vertebræ, having on either side a *venous sinus* extending along the whole length of the canal, sending a transverse communicating branch under this ligament opposite the centre of the body of each vertebra—also sending branches outwards through each intervertebral foramen. These sinuses anteriorly through the medium of the posterior petrosal sinuses are connected to the cavernous sinus.

PART VI.—SPECIAL ANATOMY.

The Back.

By dissecting the skin from the lateral thoracic incision to the middle line of the back the **panniculus** is first exposed, and then the manner in which, by uniting with the combined aponeurotic expansions of latissimus dorsi, the two small serrati (or superficialis costarum), and the dorsal trapezius, it forms the **lumbar faschia**, which extends along the whole length of the back, from the shoulders to the quarters, where it blends with the gluteal investing layer. It is attached to the superior spinous ligament and extends about six inches on either side of the spine.

Having divided panniculus and thrown it back, leaving it attached to the lumbar faschia, we expose **latissimus dorsi**, running from about the centre of the mass of muscle forming the shoulder. Its muscular portion terminates about opposite the eighth rib, where it blends with the lumbar faschia. Superiorly situated to this is **trapezius**, a triangular muscle, the apex of which is fixed to the tubercle on the spine of the scapula, its base to the superspinous ligament blending with the lumbar faschia posteriorly, anteriorly covered by levator humeri, and with it attached to the cordiform upper portion of ligamentum nuchæ. Its tendinous portion extends from its apex to its base at the withers, thus dividing the muscular portion into two parts.

The **posterior small serratus** is exposed to view by removal of panniculus. It presents a serrated margin inferiorly, its fibres running in an upward direction from the posterior margins of the seven posterior ribs to the lumbar faschia.

Below this we see the external intercostals running in a downward and backward direction from the posterior margin of one rib to the anterior margin of the next. Still lower, attached to the inferior extremity of the external surface of the ribs, is the serrated margin of obliquus abdominus externus. Having cut through and dissected back latissimus dorsi and trapezius, the operator must show the posterior attachment of **rhomboideus muscle**, which arises from the cordiform portion of the ligamentum nuchæ to within six inches of the occiput, and from the continuation of it as far back as the sixth dorsal superior spinous process, from which points it runs to become inserted into the venter surface of the cartilage of elongation of the scapula, and to that bone above serratus magnus. Now rhomboideus may be cut through, and the fore limb (if not by this time separated) may be drawn downwards, and serratus magnus traced to its inferior attachments. We now see the **small anterior serratus muscle** running from the anterior border of the ribs from the sixth to twelfth become attached by tendinous expansion superiorly to the superior spinous processes of the dorsal vertebræ from the second to the thirteenth. By dividing the dorsal and lumbar faschia and turning the serrati downwards upon their inferior attachments we expose

Transversalis costarum, a thin, complex muscle, about two inches broad, anteriorly attached to the transverse process of the last cervical vertebra, in common with longissimus dorsi, and by small tendons attached to small roughened surfaces on the posterior margin of all the ribs (in the first at the tubercle). By muscular attachments it is connected to the anterior margin of the ribs as high up as their tubercles. Posteriorly it becomes attached to the transverse processes of the anterior lumbar vertebræ. Its internal surface rests on the external intercostals, while its superior margin is in direct contact, and almost blends with

Longissimus dorsi. This is, posteriorly, an immense fleshy mass situated at the posterior part of the back and on the upper surface of the lumbar vertebræ; it is the largest muscle in the horse's body, and is attached to the crista of the ilium, extending from the antero-inferior to the supero-posterior spinous process, and occupying a slight portion of its venter surface, whereby it becomes attached to the sacro-iliac ligament and the transverse oblique and superior spinous process of the first and second sacral bones, and extends backwards into the triangular space on the supero-lateral part of the sacrum, where it is attached to the sacral and to the ilio-sacral ligaments. It is attached to the transverse, oblique, and superior spinous processes of all the lumbar and of the posterior dorsal vertebræ, to the posterior part of the heads of all the ribs, and to the transverse processes of the twelve anterior dorsal vertebræ, and the last cervical in common with transversalis costarum. At its anterior part it is separated by cervicodorsalis from complexus major; at the posterior part of the withers it comes in contact with

Spinalis dorsi, which receives from it a large and strong tendon. It is attached posteriorly to the summits of the twelve anterior superior spinous processes of the dorsal vertebræ and to the posterior margins of the superior spinous processes of the six anterior of these bones. Anteriorly with ligamentum nuchæ, it becomes attached to the superior spinous processes of the three posterior cervical vertebræ. This muscle internally lies in contact with its fellow of the opposite side, for the ligamentum nuchæ is here deficient.

The **posterior cervical artery**, after passing between the first and second ribs, and giving off the first intercostal branch, runs over spinalis dorsi to reach the external surface of ligamentum nuchæ, along which it runs up the neck parallel to the vertebral column. It arises on the right side by a root common to it, and the **dorsal artery**, from the right arteria innominata, round the common root, the recurrent or inferior laryngeal branch of the right pneumogastric nerve winds from without inwards and then runs up the neck; on the left side these two arteries arise separately, for the left recurrent nerve passes round the aorta. The dorsal artery gives off a branch which runs underneath the heads of the ribs to meet the sixth intercostal artery, the first given off by the posterior aorta. From this arch run the third, fourth, and fifth intercostals; the second intercostal is the next branch given off by the dorsal, which subsequently passes over cervico-dorsalis towards the withers. By removal of longissimus dorsi and transversalis costarum we show a series of muscles running obliquely downwards and backwards from the superior spinous processes to the oblique processes of the vertebræ, and on examination we find this arrangement obtains in the cervical lumbar and dorsal regions constituting the

Semispinales colli, dorsi, et lumborum, which commences at the superior spinous process of the dentata and extends to the sacrum. It is composed of a chain of muscular fasciculi, which run obliquely downwards and backwards from the superior spinous processes of the vertebræ to the oblique processes of the vertebræ posteriorly placed to these, sometimes passing over two or three prior to insertion. A somewhat similar muscle is

Intertransversales colli, dorsi, et lumborum, which is composed of a series of fasciculi, which in the neck are situated between the transverse processes of the vertebræ, in the back are blended with levatores costarum, and in the loins are found between the transverse processes.

Levatores costarum are a series of small triangular muscles arising from the transverse processes of the dorsal vertebræ, and running to the anterior border and external surface of the ribs as high up as their necks; superiorly they blend with intertransversales dorsi, and they are merely the superior parts of the

Intercostales externi, which consist of muscular and tendinous fibre intermixed, which run obliquely downwards and backwards from the posterior margin of one rib to the anterior margin of the rib immediately posterior to it. They extend no lower than the inferior extremities of the ribs, for between the cartilages the

Intercostales interni fill the intervening spaces with fibres running obliquely downwards and forwards. Those portions of these muscles situated between the sterno-costal cartilages (which are attached inferiorly to the lateral parts of the sternum) are termed sterno-costales externi. From this point upwards we see the internal intercostals situated internally to the external intercostals, the artery, vein, and nerve being, during a considerable part of their course, between the two muscular layers thus formed. The pleura costalis covers the inner surface of the internal intercostals.

The source of the **intercostal arteries** has been already indicated. From their origin they run over the lateral part of the bodies of the dorsal vertebræ, giving off *medullary branches* to these bones; of those intercostals which arise from the posterior aorta the right are somewhat the longest, and before reaching the intercostal spaces pass above the thoracic duct and vena azygos. In passing the intervertebral foramen, the artery gives off *a branch which passes inwards to supply the spinal cord and its membranes*. It next sends *a large branch to the muscles of the back*, and on first gaining the intercostal space is at the anterior margin of the rib behind, but soon gains the posterior margin of the rib in front, along which it passes protected by the bony groove, and by the intercostal muscles, between which it is placed. It sends ramuscules to the pleura, muscles, ribs, and skin in the neighbourhood, and the anterior intercostals help to supply serratus magnus; they are therefore the largest. At the inferior part of the intercostal spaces these vessels anastomose with the internal thoracic artery anteriorly, posteriorly with that branch which it sends along the inferior extremities of the false ribs. Thus the posterior intercostals indirectly supply the abdominal muscles and anastomose with the lumbars, circumflex of the ilium, epigastric, and abdominal branch of the internal thoracic.

The **intercostal veins**, after accompanying the arteries as far as the thorax, for the most part empty themselves into the

Vena azygos, a vessel which arises under the loins and passes with the thoracic duct and posterior aorta through the hiatus aorticus (a foramen of the diaphragm) into the thorax, where they pass in the superior part of the mediastinum underneath the bodies of the dorsal vertebræ, vena azygos being on the right, posterior aorta on the left, and the thoracic duct centrally placed. This vein passes to the anterior vena cava, with one of the branches of which it communicates by means of a subcostal branch, into which the third, fourth, and fifth intercostal veins pass, thus completing the analogy of the arrangement of the arteries and the veins of this region, for the second intercostal and subcostal veins pass to the dorsal vein, the first intercostal to the posterior cervical. The blood from the intercostal veins therefore passes to the *anterior* vena cava. The **intercostal nerves** are derived from the dorsal intervertebral nerves, each of which on emerging from the intervebral foramen, after sending fibres to and receiving fibres from the gangliated cord of the sympathetic, sends superior branches to longissimus dorsi and the other structures on the back, while its inferior branch runs between the ribs. Those most posteriorly situated are continued from the inferior

OUTLINES OF EQUINE ANATOMY.

extremities of the intercostal spaces over the external surface of transversalis abdominis, forming remarkable parallel white cords (falsely termed abdominal ribs). The most posterior of these are derived from the **lumbar nerves** which resemble the dorsal, but, of course, do not possess intercostal branches.

The **THORAX** must now be opened, which may be done by cutting through the superior and inferior extremities of the ribs from the second to about the twelfth by means of a saw, and removing the separated portion of the wall of the chest by cutting through the first and twelfth intercostal muscles. The pleura or serous lining membrane of the chest is thus exposed, and since it is very transparent we can see the structures it covers very plainly through it. The thorax is the anterior cavity of the trunk. It is pyramidal in form, its apex being anteriorly placed, its base cut off obliquely from below upwards posteriorly. Superiorly it is bounded by the under surfaces of the bodies of the dorsal vertebræ, which centrally form a prominent ridge, which anteriorly from the sixth vertebra is bounded on either side by a prominent fleshy mass, the *thoracic portion of longus colli*. On either side the **roof** of the thorax is formed by the arch of the ribs, while the straighter portion of them passes downwards to articulate with the sternum through the medium of the costal cartilages, thus with the aid of the intercostal muscles forming the **sides** of the thorax. Its **floor** is about one half the length of the roof, and is formed of the sternum with its cartilages. Its **apex** is the space between the two first ribs, through which important structures pass, *i. e.* longus colli muscles (2), common carotid, axillary (2), vertebral (2), inferior cervical (2), external pectoral (2), arteries with the corresponding veins (jugular vein corresponds to carotid artery), trachea, œsophagus, pneumogastric, recurrent, phrenic, gangliated cord of sympathetic nerves.

The **base** of the thorax is formed by the anterior, convex, surface of the diaphragm, the thin expanded muscle which separates this cavity from the abdomen. The

DIAPHRAGM presents centrally a heart-shaped tendinous portion, the fibres of which are much interlaced, and which, on the right side by separation of its fibres, forms the **foramen dextrum**, by passing through which the posterior vena cava gains the cavity of the chest, during its passage receiving blood from the flat phrenic or diaphragmatic venous sinuses. From the outer margin of this tendinous portion muscular fibres radiate towards the inner surfaces of the inferior extremities of the twelve posterior ribs, or of their cartilages, where they become inserted in forming digitations with transversalis abdominis, and towards the upper surface of the ensiform cartilage. The indentation in the upper part of the tendon is filled by three muscular masses, two of which arise from the left, the third from the right side of the tendon. Between the right muscular mass and the central one is **foramen sinistrum**, through which the œsophagus and the pneumogastric nerves pass in their course from the thorax into the abdomen; subsequently these two masses combine and form a tendon, the **right crus**, which becomes attached to the inferior vertebral ligament as far back as the fifth lumbar vertebra; the third mass terminates in a much shorter tendon is termed the **left crus** of the diaphragm, and becomes inserted into the inferior vertebral ligament below the body of the first lumbar vertebra. Between the right and left crura is situated **hiatus aorticus**, which admits the passage of the thoracic duct, posterior aorta, and vena azygos, while passing through this opening the posterior aorta gives off the *phrenic artery* which supplies the crura of the muscle. That portion of the diaphragm extending from the spine to the internal surface of the last rib passes over the anterior part of the inferior surface of the psoas muscles, and above it also we see the *gangliated cord of the sympathetic and the greater and lesser splanchnic nerves passing on either side.*

The chest is lined, and the viscera it contains covered by the **pleura**, a serous membrane arranged in such a manner as to form two sacs, one on each side of the cavity, which lie together along the middle line so as to form a septum extending in an antero-posterior direction, consisting of two adjacent layers of pleura in some parts in close apposition, in others separated by the interposition of important structures, but postero-inferiorly presenting numerous perforations, whereby one pleural sac is brought into communication with the other. This is the **mediastinum**, and extends from the internal surface of the two first ribs to the central line of the diaphragm. Centrally it contains the heart, being reflected to form the external serous layer of the pericardium. This is the *middle mediastinum*, while those portions anteriorly and posteriorly placed to the heart are the *anterior and posterior mediastina*. The **anterior mediastinum**, therefore, contains the following structures: anterior aorta and two arteriæ innominatæ with the commencement portion of the branches given off by them; the anterior vena cava with the terminating portions of some of its component veins; part of the trachea and œsophagus, pneumogastric, phrenic, recurrent, and sympathetic nerves and thoracic duct, and superiorly the terminating portion of longi colli (right and left); inferiorly, the thymus gland or its remains. The **middle mediastinum** is mainly occupied by the heart enclosed in the pericardium, and superiorly may be distinguished the large vessels which it receives or gives off (including vena azygos, which it receives indirectly), and which serve to attach it to the under surface of the spine. Also we here see the œsophagus situated on the right side of the base of the heart; the trachea terminating in bifurcating to form the right and left bronchi, and the thoracic duct winding from the right to the left side towards its termination. From the supero-lateral parts of this mediastinum also we find the visceral pleura projecting, to surround the right and left lungs. Through this mediastinum also the two pneumogastrics, the left recurrent and the two phrenic nerves, run. The **posterior mediastinum** is much longer superiorly than inferiorly, and is perforated inferiorly. At its superior part are three vessels which, after passing through or while going to hiatus aorticus, course along the under surface of the spine. On the right side vena azygos, on the left the posterior aorta; centrally the thoracic duct which seems to be simply a space between the two, so thin and uncoloured are its walls. Lower down the œsophagus accompanied by the two pneumogastric nerves courses through this mediastinum towards the foramen sinistrum, with it also runs the *left phrenic nerve*. The *right phrenic nerve* with posterior vena cava passes through a special fold of pleura situated between two lobes of the right lung.

We have now examined the pleura and have seen that it is divided into a **parietal and a visceral portion**; the former consists of the *costal* and *phrenic*, the latter of the *pulmonary and mediastinal portions*. In health these parts lie in contact with one another, the pleural sac being practically obliterated, only producing sufficient serous fluid to obviate any friction which might be caused by the movements of the parts. Thus any enlargement of the thorax causes distension of the lungs, for air rushes in through the trachea to fill up the extra space. In disease frequently large quantities of fluid collect here, and membranes are formed which, by causing union of the parietal to the visceral portion of the pleura, lead to obliteration of the sac, placing the animal in a condition analagous to that normally obtaining in birds, which class of animals is characterised by absence of pleura. Where the parietal comes into connection with the visceral pleura the bronchi and pulmonary arteries enter the lung, and the pulmonary veins emerge from it. This is termed the **root of the lung**. (The root of the left side is attached to the base of the heart.)

The **trachea** breaks up into two **bronchi**. The *right* is the largest, and passes to the right, being crossed by the œsophagus. It is larger than the left, and breaks up into three parts on joining the right lung; for this lung is larger than the left, which is deficient centrally to allow room for the movement of the apex of the heart in a direction upwards and to the left. And *it has also three lobes, while the left has but two*. Between its middle and posterior lobes is situated the pleural fold for the posterior vena cava and right phrenic nerve.

The bronchi are composed of complete cartilaginous rings; they divide and subdivide in the substance of the lung, and the cartilaginous rings become thinner and divided into several small portions until at last they disappear; so in the smaller bronchial tubes the walls are membranous. Between the cartilaginous rings and the mucous membrane are situated unstriated muscular fibres, extending in a circular manner completely round the tubes. The mucous membrane itself at first, like that of the trachea with which it is directly continuous, presents columnar ciliated epithelium; but in the smallest bronchi the cells assume the tessellated character and lose their ciliæ, so that here it resembles serous membrane. These minute bronchial tubes terminate in the **air-cells or infundibula**, elongated spaces with *saculated walls* (*air-sacs*), on the external surface of which the pulmonary capillaries are distributed in so close a network that the meshes are not so large as the vessels themselves; and to bring about more complete aëration of the blood, in consequence of the close connection of the air-cells, each layer of capillaries is in apposition on both surfaces with the thin wall of an air-cell, which has here become but little more than a layer of basement membrane covered by simple tessellated epithelium. Thus the air and the blood do not come into direct contact, but interchange of gases takes place by the process of osmosis through the layers of epithelial cells of the air-sacs and of the capillaries and their respective basement membranes. Two arteries pass to the lungs; of these the **bronchial**, derived by a root which it shares with the œsophageal, from the posterior aorta or from the sixth intercostal artery, is the nutrient vessel of the organ; while the pulmonary artery is the functional vessel, which brings that blood which requires aëration (and which will be shortly noticed). The blood obtained through these two vessels seems to become mixed in the capillaries, and returns either by the pulmonary veins into the heart, or by the bronchial veins into vena azygos, and so indirectly into the anterior vena cava. The *lymphatics of the lungs* are numerous and converge towards certain **bronchial lymphatic glands,** which may be found as grey nodular bodies at the root of the lungs. The nerves are the pneumogastric and the sympathetic, which intermix their fibres intimately in forming the **bronchial plexus at the root of the lung.** All these structures are united together by areolar tissue, to which the term *parenchyma* is applied; but this term is very indefinite, sometimes being used to indicate the whole lung-tissue as opposed to the pleura, its use therefore should be discontinued. If we now remove the exposed lung by cutting through its root we shall see the **HEART,** *in sitû.* It is enclosed in a dense membrane, the pericardium, and presents a base and an apex; and to its base, which is superiorly placed, vessels from above, from behind, and from before, are attached, and serve to sling it in position. The **pericardium** consists essentially of a membrane composed of white fibrous tissue, **fibrous pericardium,** which extends from these vessels at the base of the heart, in a downward direction, to become attached to the upper surface of the ensiform cartilage and posterior part of the sternum. It is covered externally by the layers of pleura forming the middle mediastinum, which here separate to accommodate the central organ of circulation. Internally it presents a smooth surface, for it is lined by a *serous membrane*, **serous pericardium,** which becomes reflected from the base of the heart in a downward direction over that organ, which it invests as far as the apex, and thus forms a closed sac, **pericardial sac,** which sometimes contains a little serous fluid, and allows the frictionless play of the heart on the inner surface of the fibrous pericardium, whose function is to restrain its excessive action.

If we have opened the chest on the right side we shall now be able to see the large veins which bring the blood from the various parts of the system to the right or pulmonary division of the heart.

Anterior vena cava commences at the junction of the jugulars between the two first ribs, and runs directly backwards to open into the supero-anterior part of the right side of the heart. It receives branches corresponding to the arteries given off by the arteriæ innominatæ,—axillary, vertebral, dorsal, internal thoracic, and several others, including

Vena azygos, a vein which commencing underneath the loins from the most anteriorly placed lumbar veins, runs in a forward direction underneath the spine with posterior aorta and the thoracic duct (being situated most to the right side). About opposite the sixth dorsal vertebra it dips downwards to open into the anterior vena cava just at its termination, and during its course receives the thirteen posterior intercostal veins and the subcostal branch from the dorsal; just before its termination it receives the bronchial veins. In consequence of its situation the left intercostal veins are longer than the right, but *we have observed an accessory vena azygos on the left side of the spine of the ass*. The anterior vena cava at its termination for about three inches presents red muscular fibres, so that it seems to be a continuation of the wall of the auricle; thus, in the heart removed from the body, it may be distinguished from the

Posterior vena cava, which, on emerging into the thorax through foramen dextrum enclosed in a special double fold of pleura (with the right phrenic nerve), runs direct to the postero-superior part of the right side of the heart, receiving no vessels in its course. The large arteries may be seen best by laying open the left side of the thorax. From the base of the heart, between the two ventricles the **aorta** passes upwards for about two inches. It is a large vessel with stout elastic walls, and gives off the *coronary arteries* to the substance of the heart in a manner which will be hereafter described. It terminates superiorly in the anterior and posterior aortæ. The terminal portion of the trachea lies in contact with its right side, while on the left is the pulmonary artery and also the left pneumogastric, which on passing gives off its recurrent branch, coursing from without inwards around the posterior margin of the aorta, and running forwards and up the neck to supply all the muscles of the larynx with the exception of the crico-thyroideus. In consequence of this arrangement many authorities consider that pressure upon this nerve during the heart's action, by producing atrophy of the muscles it supplies, causes that defect in respiration termed "roaring."

The **anterior aorta**, after coursing for a short distance through the anterior mediastinum, situated below and slightly to the left of the œsophagus and trachea, terminates in forming the two arteriæ innominatæ, the right of which is much the largest. The **right arteria innominata** runs in a forward direction as far as the first rib. The first branch it gives off superiorly is the *common root of the anterior dorsal and posterior cervical arteries* (which have been already noticed); around this from without inwards the recurrent branch of the right pneumogastric winds prior to passing up the neck. These vessels arise separately on the left side, for, as just noticed, the recurrent branch of that side winds round the aorta. From its anterior part the right arteria innominata gives off the vertebral, inferior cervical, axillary, and common carotid arteries. The **left arteria innominata** gives off corresponding branches with the exception of the common carotid, and from the inferior part of each vessel is given off an **internal thoracic artery**, which runs along the floor of the thorax at the lateral part of the superior surface of the sternum. It gives off branches outwards to anastomose with the external pectoral and intercostal branches, and opposite the ensiform cartilage divides, one of its branches running along the floor of the abdomen to anastomose with the epigastric artery; the other along the inferior margins of the false ribs, anastomoses with the terminal extremities of the posterior intercostal arteries, and sends branches backwards to supply the abdominal muscles. The **common carotid** is situated below the trachea in the anterior part of the thorax, between the two first ribs. It arises from the *right* arteria innominata, and terminates in bifurcating to form the right and left carotid arteries.

The **thoracic duct** after passing through hiatus aorticus runs forwards situated between vena azygos on the right and posterior aorta on the left. Opposite, about the sixth dorsal vertebra, it swerves to the left side and passes towards the anterior opening of the thorax to terminate in the left jugular or the left axillary vein. Its walls are very thin, so that its course is demonstrable but with difficulty. Almost in apposition with the inner surface of the first rib we may find the **inferior cervical ganglion**, to which the sympathetic cord running down the neck passes, after presenting a little below and in front a small nodular enlargement, the **middle cervical ganglion**. The inferior cervical ganglion also receives a band of nerve-fibres coming from the foramina in the transverse processes of the cervical vertebræ, made up of fibres derived from the cervical nerves while emerging through the intervertebral gaps in the neck. From this ganglion backwards extends the **gangliated cord**, running first against the lateral part of the longus colli muscle, then at the supero-lateral part of the thorax against the heads of the ribs, having a ganglion opposite each vertebra and being visible through the pleura costalis, by which only it is covered. Each ganglion receives fibres through the nearest intervertebral gap from the spinal cord, and sends fibres to the neighbouring intercostal nerves From the thirteen or fourteen posterior ganglia fibres are given off which combine to form the **greater splanchnic nerve**, with difficulty separable from the gangliated cord as it runs backwards, and passes above the diaphragm with the psoæ muscles to reach the semilunar ganglion in the antero-superior part of the abdomen, some of its fibres running directly to the kidney. The **lesser splanchnic nerve** arises from the two or three posterior dorsal ganglia, and accompanies the greater nerve, terminating in a similar manner. The inferior ganglion gives off fibres which run downwards to assist the pneumogastric in forming the *cardiac plexus at the base of the heart, and the pulmonary at the root of the lungs.* These plexuses have communication with the **phrenic nerve**, which arises in the neck from branches of the third, fourth, and fifth cervical nerves, communicating with the axillary plexus, and runs backwards over the base of the heart to the diaphragm, that of the left side passing through the posterior mediastinum, the right phrenic nerve accompanying the posterior vena cava in its special fold of pleura.

We can now, by division of the above-described vessels, remove the **heart** from the thorax. We find that it is a hollow muscular organ, cone-shaped, with its base upwards, having its apex directed downwards and backwards. It is covered by the visceral layer of the serous pericardium, which gives it a glistening appearance, and through which may be seen the furrows on the surface which mark its division into several cavities. Of these we see one running downwards on each side, *ventricular furrow*, one running horizontally around the base of the heart, *auriculo-ventricular furrow*. These contain the vessels of the organ with nerves embedded in a quantity of fat. The two ventricular furrows do not reach the apex of the organ, but meet on the right side a little above that point.

The heart is divided into two **sides**; *right or pulmonary heart*, *left or systemic heart*. In the fœtus these sides communicate by the *foramen ovale*, which pierces the auricular septum, being partially closed by a valvular arrangement of the lining membrane of the heart; but in the adult, under normal conditions, this becomes impervious, being still recognisable on the right side of the septum as a light circular spot, fossa ovalis, surrounded by a prominent elastic ring, annulus ovalis. Thus, after birth in mammalia, the two sides are distinct; each side consists of two **cavities**— an *auricle* superiorly, a *ventricle* inferiorly. The heart is lined by serous membrane, *endocardium*, which is continuous with the serous membrane lining the large vessels, which run to or from its cavities. On the left side, especially in the auricle, we find a layer of yellow elastic tissue between the muscular structure and endocardium. The *basement structure of the heart* is situated at the point of junction of the four cavities. Here we find a firm, irregular-shaped piece of fibro-cartilaginous substance, which in the ox is always, in the horse occasionally, ossified; from this run three yellow elastic rings; one to the right side, circumscribing the right auriculo-ventricular opening; another backwards and slightly to the left, surrounding the left auriculo-ventricular opening; the third to the left slightly forwards, from which superiorly the elastic layer of the aorta commences; while outwardly it is continued to form a ring for the commencement of the pulmonary artery (this being the weakest of them all). To these rings is attached *the muscular structure of the heart*, consisting of fibres all arranged in a more or less oblique manner. Each cavity possesses fibres peculiar to itself, while the two auricles are connected together by common fibres, and the two ventricles also by common fibres, the auricles thus being separable from the ventricles without division of muscular fibre. Thus the auricles first contract together (synchronously), then the ventricles contract simultaneously.

The muscular walls of the heart are smooth externally, but present prominences and depressions internally. The **right auricle** is situated at the supero-anterior part of the heart. It has thin walls, which collapse when the cavity is empty, and is divided into two parts, *sinus* and *appendix*. It is termed auricle from the likeness of the corresponding part in the human subject to a dog's ear; in the horse the appendix portion assumes this figure, and is inclined to the left side, where it terminates in a point. The inner surface of this part presents prominences of muscular substance arranged in a reticulated manner, *musculi pectinati*, separated by depressions, *foramina Thebesii*, into the latter small blood-vessels are supposed to open. The sinus portion presents *five openings*. At its supero-anterior part it receives the anterior vena cava, with which vessel its muscular walls are structurally continuous. At its supero-posterior part the posterior vena cava, below which is the double opening of the cardiac or coronary veins. The walls of the sinus present but few muscular prominences; a broad ridge separates the opening of the anterior from that of the posterior vena cava, this is *tuberculum Loweri*; below it may be seen the remains of the foramen ovale (fossa ovalis and annulus ovalis). In the human subject the anterior margin of this opening presents the crescent-shaped Eustachian valve; we have not seen it in the ass. Between the coronary opening and the termination of posterior vena cava is a thin double layer of endocardium, with a small quantity of muscular structure intervening, the *valve of Thebesius*. The inferior part of the auricle presents the **right auriculo-**

ventricular opening, consisting of a surrounding fibrous ring, from which hang three flaps, composed of double layers of *endocardium* (the serous membrane lining the heart, which resembles the internal coat of blood-vessels), which serve to regulate the passage of the blood from one cavity to the other. They are triangular in shape, having their outer boundary attached to the circular fibrous ring, with its terminating angles confluent with the other two sides which form the free margin to which powerful fibrous cords (*chordæ tendineæ*) are attached. These chordæ tendineæ run to large projections from the inner surface of the muscular wall of the ventricle, which receive the special name *musculi papillares*; they are three in number, one projecting from the outer wall, two from the septum ventriculorum. From one musculus papillaris chordæ tendineæ run to each of the three flaps of the tricuspid valve.

The wall of the **right ventricle** is thicker than that of the auricle, and presents on its internal surface three forms of muscular prominences (**carneæ columnæ**). The *musculi papillares*, having one end free, constitute the first set; the *moderator bands* run from one side of the cavity to the other, often consist simply of yellow elastic tissue covered by serous membrane, they form the second; while the *third or true carneæ columnæ* are attached by both ends and by one margin, thus resembling the musculi pectinati of the auricle. The *right ventricle* does not extend to the apex of the heart, which is wholly formed by the left, but superiorly it forms a funnel-shaped cavity (termed conus arteriosus), behind the posterior flap of the tricuspid valve, from the upper extremity of which the pulmonary artery commences. The communication of the ventricle with this vessel is guarded by three crescentic flaps which constitute the *semilunar pulmonary valves*. They have a free and an attached margin. The free margin centrally presents a point on which we sometimes see a small nodule (*corpusculum arantii*), which serves to ensure complete closure of the opening by the valves. The attached margin is continuous with the pulmonary fibrous ring and its extremities meet the other flaps. The **pulmonary artery** runs backwards and in passing under the posterior aorta is connected to it by *an impervious cord, the remains of the ductus arteriosus of the fœtus*. After this it bifurcates, its larger division passing to the right lung, its smaller to the left. On examining the inner surface of this vessel three saccular dilatations (*sinus Valsalvæ*) of its walls may be noticed behind the valves, serving to accommodate them during the rush of blood from the ventricle, and to allow blood to collect behind the flaps in sufficient quantity when this rush ceases to throw the valves into position and thus prevent regurgitation of blood. The pulmonary arteries divide and subdivide in the lungs, and finally terminate in capillaries in which the blood is oxygenated, and from which it is carried by the pulmonary veins to the left auricle of the heart. The **pulmonary veins** unite and reunite until by (generally) four openings they pour their contents into the heart; they never leave the lung substance, for it is here also attached to the heart substance. These veins have no valves. The **left auricle** very much resembles the right, from which, however, it differs in the greater amount of yellow elastic tissue lining it, and in presenting more marked musculi pectinati and foramina Thebesii, but no fossa ovalis, annulus ovalis, tubercrum Loweri, or Thebesian valve. Its **auriculo-ventricular opening** too is guarded by a stout valve consisting of but two flaps, *the bicuspid or mitral valve*. It leads into the **left ventricle**, similar in structure and figure to the right ventricle, but with only two musculi papillares, both of which are attached to the outer wall of the cavity. Its moderator bands generally contain muscular fibres, and its walls consist of very thick muscular substance. This compartment, however, at the extreme inferior part, where it alone forms the apex of the heart, consists of little more than a double layer of serous membrane, the endocardium lying in contact with the pericardium. The thickness of the muscular wall is related to the force required in driving the blood through the system when the elasticity of the coats of many large and strong vessels has to be temporarily counteracted. The blood from this cavity passes in an upward direction into the **aorta**, which is situated on the right side of the pulmonary artery, to which one of its primary divisions (posterior aorta) is connected by an elastic ligament. The opening into this vessel, like that of the pulmonary artery, presents *semilunar valves*, but they are larger, stronger, and present more marked corpora arantii; while the sinus Valsalvæ into which they fit are deeper than those of the pulmonary artery, so that the vessel is largest and its walls thinnest and most liable to rupture here. From the posterior and also from the internal sinus Valsalvæ a coronary artery arises, going to supply the substance of the heart. The **left coronary artery**, the shortest, runs directly down the left ventricular furrow, and its terminal branches anastomose with those of the right at the point of union of the two ventricular furrows. It gives off a branch in a posterior direction which passes in the posterior auriculo-ventricular furrow to the vessel of the right side. The **right coronary artery**, after winding through the anterior auriculo-ventricular furrow, meets the posterior branch of the left artery and then runs down the right ventricular furrow to meet the left artery at the inferior extremity. The blood from the heart is returned by the **coronary or cardiac veins**, of which the left is the longest; they terminate almost by a common opening in the right auricle as already described.

The heart is supplied with nerve power from three sources: By the *pneumogastric* and *gangliated cord of the sympathetic* through the medium of the cardiac plexus, and by *small sympathetic ganglia embedded in its substance*. The latter endow it with its regularity of contraction (rhythmic action), the pneumogastric serves to restrain, the gangliated cord of the sympathetic to stimulate its action.

The muscular fibres of the heart are involuntary; they nevertheless present slight striation and a modena red colour, similar to that of voluntary muscle. They differ from the latter, however, in the imperfection of their striation, in the absence of a sarcolemmatous sheath, and in frequent anastomosis of the constituent fibrillæ.

In the infero-anterior part of the thorax, at the lower part of the anterior mediastinum, especially developed in the fœtus, in whom it extends a short way up the neck, is the **thymus gland**, a body resembling a salivary gland in appearance, and pouring the product of its lobules into a central cavity, from which, as there is no duct, it seems to be reabsorbed. It is a vascular gland, and is the true "sweetbread."

PART VII.—SPECIAL ANATOMY.

The Fore Extremity.

THE skin having been dissected back from the withers (a work rendered easy by the primary incisions on either side) down to a little below the elbow-joint, the **panniculus** is shown extending over the limb on the outside as far as this joint, becoming indirectly attached to the spine of the scapula, the point of the shoulder, the humeral ridge, &c. Anteriorly it is somewhat intimately connected with levator humeri, posteriorly it covers the muscles of the shoulder, and dorsal trapezius, and extends on to the lateral part of the thorax. From the lateral parts of the thorax some fibres of panniculus run forwards to the internal surface of the shoulder. The panniculus having been removed, **levator humeri** must be traced in its attachments, to the spine of the scapula (blending with trapezius), the external ridge of the humerus, with the fibrous band running downwards from it, and the antero-inferior part of the humerus. The latter attachment being also that of **pectoralis transversus**, which will be found running from the under surface of the sternum to this attachment, and by a broader part (almost separated from the former) to the faschia covering the inner surface of the arm. By division of levator humeri along the anterior margin of the scapula, **pectoralis anticus** is exposed; it is a triangular elongated muscle arising from the lateral part of the cariniform cartilage and from the external surface of the anterior sterno-costal cartilages and muscles, and running along the front of the shoulder, passing over pectoralis magnus and growing thinner superiorly to blend with the faschia covering antea spinatus and to reach the spine of the scapula. Between this muscle and those of the neck is situated a quantity of loose cellular tissue in which much fat and the **prescapular lymphatic glands** are situated. The **subscapulo-hyoideus** muscle runs on the inner side of pectoralis anticus, forming a broad muscular layer attached to the tendinous aponeurosis of subscapularis; in proceeding up the neck about the middle third this muscle becomes intimately connected with the internal surface of levator humeri, so that these muscles are not easily separable; it subsequently passes between the jugular vein and the carotid artery to its insertion into the hyoid bone.

The muscles connecting the fore limb to the various neighbouring parts must now be divided (reference should be made to the several regions under the heads of which their attachments are described), and the *external scapular muscles* exposed, these are—

Antea spinatus, running from the anterior dorsal fossa of the scapula, by a rounded prominent border clothing the major portion of the anterior costa of that bone and becoming attached to a ligament, which runs from here to the coracoid process, between which and the bone pass the dorsal scapular vessels and nerves. Inferiorly it bifurcates and with pectoralis magnus becomes inserted into the external and internal tubercles on the anterior part of the humerus; the surface between the two heads is lined by a tendinous theca or sheath, which is reflected on the middle tubercle of the bone to facilitate the gliding of the cartilaginous superior extremity of

Flexor brachii or coraco-radialis, which having arisen by a strong tendon from the coracoid process of the scapula, after passing through this sheath, where its tendon is indented to fit upon the middle tubercle, runs down the front of the humerus inclined to the inner side, forming a prominent muscular cord which sends a strong band of white fibrous tissue downwards to extensor metacarpi magnus, and becomes inserted into the roughened spot previously noticed as situated at the supero-internal part of the anterior surface of the radius and into the capsular ligament of the elbow-joint; at the antero-superior part of its tendon we see some muscular fibres.

Postea spinatus is to be found in the posterior fossa of the dorsum of the scapula, which it occupies for its superior two thirds; separating below it forms two tendons, one of which is attached to the fourth or external tubercle of the head of the humerus, while the other plays over this, from which it is separated by a synovial bursa, and becomes inserted into the upper part of the ridge running down from the external tubercle. The inner surface of the first head is attached to the capsular ligament of the shoulder-joint.

Teres externus arises from faschia which covers postea spinatus and extends as far forwards as the spine of the scapula, also from the superior part of the posterior costa of the scapula. It is at first composed of two layers, but these combine inferiorly to become inserted into the lower part of the humeral ridge. Between this tendon and the lower tendon of postea spinatus the

Scapulo-humeralis externus is inserted, after passing over the capsular ligament of the shoulder-joint and becoming attached to it, and arising from a roughened ridge at the postero-external part of the neck of the scapula at the lower part of the fossa postea spinatus in front of caput magnum, extending to a slight extent around its inferior extremity to its posterior surface. By removing the superior attachment of this muscle, we generally expose a branch from the posterior scapular artery passing through the medullary foramen of the scapula.

Underneath the above-named muscles to the posterior surface of the humeral ridge the

Caput medium of the **triceps extensor brachii** is attached. The triceps is that large mass of muscle which fills up the triangular space between the scapula and the humerus; as its name indicates, it consists of three parts or heads, named according to their respective sizes. **Caput magnum** is attached superiorly along the major portion of the posterior costa of the scapula, not extending, however, to the extreme inferior part. It is attached to the capsular ligament of the shoulder-joint and covers scapulo-humeralis posticus, and at the extreme inferior part becomes large and thick, and internally to its superior attachment presents a groove in which rests the teres major vel internus; between the two muscles is the **posterior scapular artery** with its accompanying vein and nerve. It is a branch of the axillary artery and runs upwards, sending branches to the muscles, between which it runs to subscapularis. The fibres of this head converge, and run towards the point of the ulna, into which they become inserted through the medium of a short stout tendon which passes over a bursa prior to insertion. **Caput medium** runs from the posterior part of the humeral ridge to be attached by a wide insertion to the tendon of the caput magnum and to the outer surface of the olecranon. **Caput parvum** arises from the upper part of the internal ridge which bounds the olecranian fossa of the humerus and passes to the point of the ulna and to the internal surface of its olecranian process. The internal surface of this head is in contact with the humeral vessels and nerves derived from the axillary plexus; also above with the humeral attachments of

Teres internus, which arises from the superior part of the posterior costa of the scapula, is externally in contact with caput magnum as above described, and inferiorly terminates by a tendon common also to latissimus dorsi, at a roughened spot on the internal surface of the middle third of the humerus; over this tendon the humeral artery and vein and several important branches of the axillary plexus of nerves pass. It will be seen that this muscle exactly corresponds in its course to the teres externus on the outside of the limb. To the latissimus dorsi over its inferior extremity runs the external thoracic branch of artery.

On looking at the external surface of the shoulder, we see that the venter surface of the scapula is covered over its inferior two thirds by

Subscapularis, which is covered by faschia, from which subscapulo-hyoideus arises. Superiorly it presents a margin of attachment divided into three projections, whereby it leaves two triangular spaces for insertion of serratus magnus. Inferiorly it becomes more narrow, and finally for the most part tendinous, and inserted into the tubercle situated on the inner surface of the head of the humerus. The anterior margin of this muscle is in contact with antea spinatus, which here presents a rounded surface apt to be mistaken for another muscle until the limb is turned over; between these two muscles, above the coracoid process, the anterior scapular artery (a branch of the axillary) passes with its accompanying vein and nerve.

Attached to the process on the internal surface of the coracoid process is the tendon of

Coraco-humeralis, which, emerging from between the two muscles, passes over the insertion of subscapularis, over which it plays through the medium of a bursa, then breaks up into two muscular portions which go to be attached, the superior one to the internal surface of the superior third of the humerus, the inferior to the superior part of the inferior third of the anterior surface of that bone. Between these two heads runs a branch of the humeral artery (with its vein and nerve) to supply flexor brachii. The axillary nerves and vessels will be found in contact with the upper part of this muscle.

Scapulo-ulnaris runs from the posterior angle of the scapula with caput magnum, and covering the internal surface of the posterior part of the triceps, expands below to become inserted into the inner surface of the olecranon and the faschia covering the internal surface of the arm, forming an aponeurotic expansion which extends upwards to blend with its superior attachment.

By removal of triceps we expose two muscles, both of which are attached to the capsular ligaments of the joints with which they respectively are in contact.

Scapulo-humeralis posticus consists of a few fleshy fasciculi, which run from immediately behind the glenoid cavity of the scapula, emerging from within the attachment of caput magnum, and after passing over and becoming attached to the capsular ligament of the shoulder-joint, dip between the fibres of humeralis externus to become inserted to the extreme postero-superior part of the humerus immediately below the articular cartilage of the head.

Anconeus arises from the superior part of the ridges which bound the olecranian fossa of the humerus (especially from the outer), becomes attached to the capsular ligament of the elbow-joint, which extends around the articular part of the fossa, and becomes inserted into the antero-superior, inclined to the outer part of the olecranon.

Humeralis externus arises from the posterior part of the humerus immediately beneath its head, where it covers scapulo-humeralis posticus, and from the back of the bone as far down as the supra-condyloid ridges. It winds around the external surface of the bone in a spiral fossa bounded by ridges (with the spiral nerve), passes beneath the band of white fibrous tissue, mixed with yellow elastic, which runs from the inferior extremity of the humeral ridge to the external ridge of the inferior extremity of the humerus, passes between flexor brachii and extensor metacarpi magnus beneath the tendinous band connecting these two muscles, and having become attached to the anterior part of the capsular ligament of the elbow-joint, forms a flat tendon which passes underneath the inferior portion of the internal lateral ligament, where it is lubricated with synovia, and becomes inserted into the internal surface of the head of the radius and into the inner sharpened border of the ulna.

Muscles attached to the capsular ligament of the scapulo-humeral joint:

Antea spinatus, postea spinatus, subscapularis, scapulo-humerales externus and posticus, and caput magnum. Flexor brachii is separated from it by a quantity of fat and its bursa, coraco-humeralis by the subscapularis tendon and a bursa.

The **axillary artery** must now be viewed as a whole. It commences from the anterior part of arteria innominata within the chest, and takes a turn around the anterior margin of the first rib below the inferior head of scalenus at its posterior attachment. It thus gains the external part of the thorax and the internal surface of the limb, so that on raising the limb after division of the pectoral muscles we see this vessel with its corresponding vein, and the axillary plexus of nerves, with the **axillary lymphatic glands** embedded in loose cellular tissue, in the neighbourhood of the shoulder-joint. It gives off arteria dorsalis scapulæ (or anterior scapular artery), posterior scapular (or subscapular), external thoracic, and humeral arteries, the course of which we have already noticed; also the **humeral thoracic** which runs to supply pectoralis magnus and antea spinatus at the point of the shoulder. The **humeral artery** in its course along the internal surface of the humerus gives off *a branch anteriorly, which passes between the two heads of coraco-humeralis to flexor brachii, a large branch backwards to triceps extensor brachii*, and *a branch inwards* opposite the superior part of the inferior third of the bone, which pierces its substance by *passing through the medullary foramen*. The artery then passes to the anterior margin of the bone and breaks up into the ulnar, spiral, and radial arteries. The **ulnar** runs backwards, and very soon breaks up to form a number of smaller vessels, which all run to structures near the olecranon. The **spiral** passes over the front of the elbow-joint, and runs down the anterior surface of the radius in company with a vein and nerve. The **radial**, after passing over the internal surface of the elbow-joint, dips underneath flexor metacarpi internus, on the inner surface of which it runs, in company with an artery, vein, and nerve, as far as the knee, sending out branches to the flexors in its course, even to *flexor metacarpi externus; the branch to this muscle runs through the arch in the union of the ulna to the radius, and in the passage gives off the medullary artery of the radius.*

The **axillary plexus of nerves** is formed by the union of nerve-fibres from the three posterior cervical and the first dorsal intervertebral nerves, which also receive a branch from the sympathetic system. Emerging between the long and short heads of scalenus, the plexus is thus separated in winding around the first rib from the artery and vein by the lower head of that muscle. From this point collections of fibres pass in company with the corresponding arteries, thus we find the *dorsalis scapulæ, humeral thoracic, external thoracic, and subscapular nerves,* which differ only in arrangement from the arteries in that their fibres separate before arriving at the parts which they supply, and become thus more diffused over its surface. In addition to these the axillary plexus gives off the *ulnar, spiral and radial nerves, for we find no recognised humeral nerve.* The **radial** runs with the artery of the same name, and below the knee forms the internal metacarpal nerve; it gives off a branch which runs forwards with the spiral artery to the front of the elbow-joint, where it meets the true **spiral nerve**, which ran in a backward direction, behind the humerus, piercing the fleshy mass of the shoulder by passing between caput parvum and caput magnum to gain the inferior margin of humeralis externus, with which it winds round the external surface of the bone, gains the anterior surface of the joint, and meets the artery, with which it runs to the extensor muscles and the skin covering the front of the limb. The **ulnar nerve**, from the point of the ulna, sends a *long branch down behind the limb underneath ulnaris accessorius*, which below the knee becomes the external metacarpal nerve.

Two important veins are to be found in the region of the shoulder. The **humeral vein** receives blood from the radial and ulnar veins, and also from a branch which runs to it from just in front of the termination inferiorly of flexor brachii. It then runs up the limb posteriorly placed to the artery and terminates in the **axillary vein**, which in most respects resembles the artery and opens into the anterior vena cava. Along the internal surface of the fore limb, in the living animal, we may see a large vein running upwards along the subcutaneous portion of the radius. This is the **superficial brachial, cephalic, or plat vein,** from which blood is sometimes extracted by the process of venesection; at the insertion of flexor brachii it sends a branch internally to that muscle to join the humeral vein as above mentioned, but the main portion runs along the groove between this muscle and extensor metacarpi magnus, and proceeding upwards empties its contents into the jugular vein; opposite the front of the elbow-joint this vessel receives the **spiral vein.** Now the skin may be removed from the internal surface of the arm; this may be done by an incision along the central line from the elbow-joint to the lower part of the knee, through the circular horny process or **chestnut** found opposite the inferior part of the superior third of the radius, and dissecting the skin backwards and forwards. It will be advisable also to remove the integuments from the outer surface of the limb as low as the knee, if sufficient time is available for examination of both aspects of the arm (otherwise the white fibrous tissue of the outside will become dry and hard by exposure to the air). Thus the aponeurotic expansion of pectoralis transversus and panniculus is exposed, it inferiorly gradually blends with the ordinary subcutaneous areolar tissue. By division of it we bring to view the **faschia of the arm,** a dense layer of white fibrous tissue running round the back of the limb from the outer sharpened border of the ulna to the inner surface of the radius, supero-internally it terminates by blending with the internal lateral ligament of the elbow-joint, and by becoming attached to the point of the ulna, between these two attachments it receives the insertion of scapulo-ulnaris. Inferiorly it blends with the annular ligament of the knee, which is nothing more than a thickening of this faschia in consequence of the addition to its substance of fibres, some of which run from the inner surface of the radius just above the knee to trapezium, others from the inner surface of the knee obliquely upwards and outwards. From the internal surface of the annular ligament of the knee thick fibrous bands pass to enclose the tendons situated at the back of the knee-joint. The first muscle posteriorly situated to the radius is—

Flexor metacarpi internus. It arises from the internal surface of the inferior extremity of the humerus, where it is attached to a slightly prominent ridge, passes over the capsular ligament of the elbow-joint, to which it is firmly attached, and runs downwards along the internal margin of the posterior surface of the radius covering the radial artery, vein, and nerve. It then runs through a synovial sheath behind the knee, and becomes inserted into the head of the inner small metacarpal bone. This sheath is strengthened externally by a fibrous band reflected from the internal surface of the annular ligament, which superiorly blends with the *superior suspensory ligament,* a dense white fibrous tissue band running from the posterior part of the inferior extremity of the radius to the perforans and the perforatus tendons. This septum is broad superiorly, but inferiorly grows narrow, and separates the tendon of flexor metacarpi internus from those of the perforatus and the perforans. Opposite the superior part of this reflection a small branch runs from the radial artery to become superficially placed on the inner surface of the limb, the *small metacarpal,* which anastomoses with the recurrent branch of the large metacarpal, and gives off the medullary artery of the large metacarpal bone. The **annular ligament of the knee** terminates below in a definite band of white fibrous tissue at the inferior part of the middle third of the internal surface of the inner small metacarpal bone; between this band and the remainder of the ligament the *inner metacarpal vein* passes to gain the cutaneous surface of the limb, and from it the superficial radial vein arises, which passes upwards to the superficial brachial vein.

Flexor metacarpi medius is situated posteriorly to flexor metacarpi internus. It is broad superiorly, where we see some of its tendinous fibres running transversely from the point of the ulna to get a firm insertion into the capsular ligament of the elbow-joint, and into the inferior part of the ridge running upwards from the inner condyle of the humerus. Under this part run the ulnar artery, vein, and nerve. The **ulnar nerve** proceeds down the back of the limb at first underneath ulnaris accessorius, and then about opposite the inferior part of the middle third of the radius passes under the tendon of that muscle to gain the inner surface of flexor metacarpi externus. The postero-superior part of flexor metacarpi medius blends with ulnaris accessorius, becoming attached firmly to the posterior surface of the superior extremity of the olecranon; inferiorly it blends with the faschia of the arm and with flexor metacarpi internus, the tendinous band so formed becoming firmly attached to the superior margin of trapezium, and to the annular ligament of the knee.

Flexores pedis perforans and perforatus.—Attached to the infero-posterior part of the ridge running upwards from the inner condyle of the humerus and to the capsular ligament of the elbow-joint, are numerous muscular fasciculi, comprised mainly of four masses which run down the posterior surface of the radius and inferiorly combine, three of them forming the perforans tendon, the other the perforatus; the first is triangular in shape, being concave on its posterior surface for the reception of the second. To the inner margin of each tendon runs a part of the superior suspensory ligament, that to perforatus being the widest and strongest. Here the tendons of radialis and ulnaris accessorius blend with that of perforans. The tendons of perforans and perforatus with the metacarpal artery, vein, and nerves run together in a synovial sheath, separated from that of flexor metacarpi internus by a fibrous layer from the annular ligament. This sheath superiorly is bounded by the superior suspensory ligament, extends downwards for about four inches, terminating inferiorly opposite the heads of the small metacarpal bones, where it is bounded by the **subcarpal ligament,** a diffused layer of ligamentous fibres which arises from the inferior part of the annular ligament, where it immediately invests the posterior surface of the small bones of the knee, and runs mainly to the perforans tendon, though a few of its fibres run on either side of that tendon to perforatus. Perforans is said to form a sheath for perforatus at the back of the knee, this is hardly correct; it forms the anterior part of an imperfect sheath which is laterally composed of superior suspensory ligament, and another layer of the annular ligament of the knee, but between the two tendons is a synovial sheath, which on the outer side communicates with the former sheath, but does not extend so far inferiorly (by one and a half inches), though superiorly it extends as far up as the superior part of the inferior third of the radius. We must temporarily defer the examination of the remaining portion of these tendons.

THE FORE EXTREMITY.   211

Radialis accessorius arises from the posterior surface of the middle third of the humerus inclined to its outer side; its fibres become inserted into the perforans tendon, in blending with the superior suspensory ligament above and behind the knee.

Ulnaris accessorius runs to the same insertion, after arising from the posterior part of the olecranon in common with flexor metacarpi medius. Its muscular portion covers the ulnar nerve and is covered by the faschia of the arm.

On examination of the external surface of the arm we find that the **faschia of the arm** blends anteriorly with the fibrous band of extensor metacarpi magnus, which runs from flexor brachii. Supero-externally its fibres converge to form a white tendinous band (which contains some yellow elastic fibres, and to which levator humeri and pectoralis transversus are attached). This, superiorly, is firmly attached to the inferior extremity of the ridge running downwards from the external tubercle of the humerus, passes over and binds down humeralis externus in its course over the external surface of the humerus, and posteriorly runs to become attached to the ridge passing upwards from the external condyle of the humerus, inferiorly it blends with caput medium.

Extensor metacarpi magnus at its superior attachment to the ridge running upwards from the external condyle of the humerus, blends with extensor pedis. It is also attached by a strong tendon to the anterior part of the capsular ligament of the elbow-joint, thus separating humeralis externus from that ligament. Between the upper part of this muscle and humeralis externus runs the spiral nerve, which after meeting the spiral artery and vein and their small accompanying nerve at the front of the joint proceeds with them down the front of the limb under the muscles. At the upper part of the front of the knee its tendon is in contact with extensor metacarpi obliquus, which passes over it in its course from without inwards and downwards, and is separated from it by a bursa. Passing then through the central one of the three grooves at the anterior part of the inferior extremity of the radius, and through a special channel in the annular ligament of the knee, where it is lubricated with synovia, the tendon of this muscle becomes inserted into the prominence on the antero-superior part of the large metacarpal bone, slightly inclined to the inner side.

Extensor metacarpi obliquus arises from the outer part of the middle third of the radius between extensor suffraginis and extensor pedis. Its muscular fibres in running under the tendon of extensor pedis converge, and they form a tendon which runs over extensor metacarpi magnus (from which it is separated by a bursa), passes through the oblique groove on the inner surface of the inferior extremity of the radius, through a channel in the annular ligament of the knee, lubricated by synovia, to become inserted into the head of the inner small metacarpal bone through the medium of the inner lateral ligament of the knee.

Extensor pedis has a penniform arrangement. Superiorly it blends with extensor metacarpi magnus in becoming attached to the external condyle of the humerus, and its fibres are attached to the inner lateral and capsular ligaments of the elbow-joint, and to the outer part of the radius as far down as the inferior part of the superior third. A remarkable round band of its muscular fibres runs underneath extensor suffraginis to the foramen between the ulna and the radius, where it becomes attached in contact with the branches of the radial vessels which pass through the foramen. While passing over extensor metacarpi obliquus the muscle becomes tendinous and passes through the antero-external groove at the antero-inferior part of the radius and the annular ligament in a special channel, lubricated with synovia. *Below the knee it sends a branch to assist in forming ligamentum extensorum.* Between the upper part of this muscle and extensor metacarpi magnus is a slight inflection of the faschia of the arm, but a much thicker fold separates it from

Extensor suffraginis, and forms a sheath in which this muscle plays, and whereby from the elbow to the knee it is firmly fixed to the lateral part of the radius. It arises superiorly through the medium of the external lateral ligament from the outer condyle of the humerus, at first forms a bridge over the erratic portion of extensor pedis and then becomes fixed to the radius as above mentioned. It passes through the external groove on the antero-inferior part of the radius, and through the annular ligament, from which it receives a fibrous cord opposite the head of the small metacarpal bone, and below the knee becomes rather expanded on the external surface of the large metacarpal bone, receives a portion of tendon from extensor pedis, and is here termed **ligamentum extensorum.** After passing over the capsular ligament of the fetlock-joint to which it is attached, it becomes inserted into the supero-anterior and external part of os suffraginis.

Flexor metacarpi externus, superiorly, is attached to the extreme posterior part of the ridge running from the external condyle of the humerus and to the capsular ligament of the elbow-joint (being the only flexor of the forearm which arises from the outer surface of the limb) ; it runs downwards and inferiorly blends with the fibrous sheath of extensor os suffraginis (which is here part of the annular ligament of the knee), becoming inserted into the superior margin of the trapezium, while it sends a small round tendon through the groove running obliquely downwards and forwards on the external surface of this bone (which is lubricated with synovia), and passes to the head of the outer splint bone.

Remove the skin from the remainder of the limb as far as the hoof, clean the tendons and ligaments thus exposed by careful removal of the areolar tissue anteriorly; the tendon of extensor os suffraginis will be found arranged as above indicated; that of extensor pedis, after playing over a large bursa at the antero-inferior part of the large metacarpal bone, grows wider in becoming attached to the front of the capsular ligament of the fetlock, receives on each side a strong band running from the inferior extremity of the superior sesamoideal ligament obliquely over the lateral parts of the pastern bone, and becomes inserted into a somewhat heart-shaped prominence on the anterior surface of the coronal process of os pedis. It is firmly fixed to the anterior surfaces of ossa coronæ and suffraginis, and of the capsular ligaments of the pastern and coffin joints by dense white fibrous tissue, and in the same way is united on both sides of its extreme inferior part to the lateral cartilages.

From the inferior margin of trapezium a portion of the annular ligament runs to the supero-lateral part of the small external metacarpal bone ; it serves to protect the nerves and blood-vessels of this neighbourhood.

The radial artery breaks up into two vessels just at the superior suspensory ligament, the smaller, **small metacarpal** artery, passes over the annular ligament of the knee and subsequently becomes internally placed to the inner splint bone. At about the inferior part of the superior third of the posterior surface of the large metacarpal bone it gives off *the medullary artery of that bone*, and about at the bifurcation of the superior sesamoideal ligament it anastomoses with *a small recurrent branch of the* **large metacarpal artery**, which we last noticed as being enclosed with the nerves in the synovial sheath of the flexor tendons at the back of the knee. It (the large metacarpal) passes down the back of the leg in contact with the inner margin of these tendons, and subsequently between them and the superior sesamoideal ligament, where it gives off the recurrent branch, and then bifurcates to form the **plantar arteries.** If we trace out one of these arteries we shall find that it passes on to the lateral part of the fetlock, where it is subcutaneous and where the corresponding vein is in front, the nerve behind it. It gives off a large branch forwards, ramifying in front of os suffraginis, and small unnamed *branches backwards to the heel*, and then runs directly to the internal surface of the lateral cartilage. Just at the lateral part of the coffin-joint it gives two branches forwards and one backwards. The two running forwards anastomose with their fellows from the opposite side at the front of the coronet, one externally situated to the extensor pedis tendon being the *superficial coronary* and sending on either side of the tendon a branch inwards to the other, which being internally placed to the tendon is termed the *deep-seated coronary artery*. The posterior *branch goes to the frog* and bifurcates, one of its branches running to the bulb, one to the apex of that organ, each of them anastomoses with its fellow. The plantars give off the **lateral laminal arteries,** while running over the internal surface of the lateral cartilages, and these pass outwards through the foramen in the wing of the os pedis (or between os pedis and the cartilage), running forwards in a groove to anastomose at the front of the bone with each other, and giving branches superiorly to the coronary arteries, inferiorly to the circumflex artery of the toe, and inwards to the arterial circle formed in the substance of os pedis by the anastomosis of the two plantars, which, after passing through grooves on the under surface of os pedis between the inferior broad ligament of the navicular bone and the perforans tendon, enter the bone through the foramina on its plantar surface. Branches running downwards through the substance of the bone from this **circulus arteriosus of the foot** pass out of the large foramina at the outer margin of the plantar surface of os pedis, where they form a continuous arterial circle, the **circumflex artery of the toe**, which receives branches superiorly from the lateral laminal arteries, which sometimes mainly form it; and on the plantar surface some large arteries (*solar*) radiate to it from the artery of the apex of the frog. It will be thus seen that the arteries form an intricate plexus, not only in the sensitive structures, but also in the bones of the foot; the **veins** are more numerous, they form an *intraosseous plexus*, the vessels of which mainly converge to a trunk corresponding to the plantar artery inside the lateral cartilage; and also a *superficial plexus* composed of vessels with but small meshes, which are found in the whole of the vascular investing and secreting layer of the foot, they converge towards certain larger veins which are on the external surface of the lateral cartilages and form a vein which unites above the cartilage with the intraosseous vein to form the plantar which runs upwards (receiving branches), with and in front of the artery to unite with its fellow above the fetlock. From this junction four or five vessels run upwards, being the satellites of the large and small metacarpal arteries, and form the radial and superficial brachial veins respectively; since the smaller veins of the foot have no valves, they can be injected from the plantar.

After passing with the flexor tendons through the annular ligament of the knee, the metacarpal nerves accompany the tendons as far as opposite the superior part of the inferior third of the large metacarpal bone. The **external metacarpal nerve** is derived from the ulnar, the **internal metacarpal nerve** from the radial. The internal gives off fibres which form a trunk extending obliquely downwards and outwards between the skin and the flexor tendons to the external metacarpal nerve. At the side of the fetlock each nerve divides into three parts; one runs over the front of os suffraginis; it is separated by the plantar vein from the middle branch which runs to the coronet, and is separated by the plantar artery from the *posterior plantar nerve* which supplies the major portion of the foot with nerve force.

PART VIII.—SPECIAL ANATOMY.

The Abdomen.

THE operator commences by dissecting the skin downwards from the middle thoracic incision towards the middle line of the abdomen. Thus he exposes the **panniculus**, extending over the lateral part of the belly from the external surface of the shoulder. Infero-anteriorly is situated *pectoralis magnus*, which is superiorly slightly covered by panniculus; it is about five inches wide, and covers about one third of the length of the under surface of the belly. Posteriorly the panniculus divides into two layers, one of which covers the outer surface of the quarter, the other runs to the inner surface and, as these two layers are intimately connected, a somewhat tense sheath is formed, which is superiorly attached to the antero-inferior spinous process of the ilium, inferiorly around the patella. It blends posteriorly with the faschia of the quarter. It is the portion of panniculus extending to this sheath in the ox, which, covered by skin, forms that thin fold termed by butchers "the flank." Below this is the **superficial abdominal faschia**, a smooth yellow layer formed of elastic tissue which is immediately subcutaneous. It is widely spread over the aponeurotic expansions of the abdominal muscles as far forwards as the ensiform cartilage, backwards as the symphysis pubis. Thick towards the linea alba, gradually grows thinner on approaching the muscular portion of obliquus abdominis externus, and at the postero-inferior part sends a process which, in the male, combines with that of the opposite side, forming the *suspensory ligament of the penis*, in the female the *elastic investment which covers the mammary gland*, and which sends trabeculæ inwards to separate its lobes. Internally this layer is attached to the external oblique abdominal muscle, externally it affords attachment to pectoralis magnus anteriorly, posteriorly it becomes inflected to the internal surface of the thigh.

THE ABDOMEN. 217

The fore limb should now be elevated by means of an iron fork, and the pectoral muscles dissected. **Pectoralis transversus** is superficially situated, it blends with its fellow at the under part of the cariniform cartilage, and anterior part of the sternum, being connected to these firmly at its anterior part by areolar tissue; posteriorly, as far as the lower extremity of the fifth rib, so that it has free play over this surface; from this its fibres run to the inner side of the fore limb, and with levator humeri become attached to the fibrous band running from the inferior extremities of the humeral ridge to the infero-external part of the humerus.

Anteriorly to the firm attachment of pectoralis transversus is the originating point of *sterno-maxillaris*, above which *sterno-thyro-hyoideus* is attached to the cariniform cartilage. Higher up **pectoralis anticus** is attached to the lateral part of the sternum, extending as far back as about the fourth sternal bone, and being slightly united to the corresponding sterno-costal cartilages; it runs upwards to the antero-superior part of the scapula.

Pectoralis magnus runs from the postero-lateral part of the sternum, and from the under surface of the ensiform cartilage and from the anterior part of faschia superficialis abdominis to the external and internal tubercles upon the anterior part of the head of the humerus, where it is attached with antea spinatus, and between which attachments runs the superior cartilaginous tendon of flexor brachii.

By section of the pectoral muscles a quantity of loose areolar tissue is shown, into the inferior part of which **subscapulo-hyoideus** seems to be attached, gaining apparently a very unsubstantial origin. The axillary artery may now be seen running to the internal surface of the shoulder-joint, while the axillary plexus of nerves is seen sending off numerous trunks. The **axillary artery** after being given off by the arteria innominata, winds round a smooth surface on the anterior border of the first rib below the inferior attachment of the scalenus muscle; on gaining the internal surface of the shoulder-joint it breaks up, sending the **humeral thoracic** branch anteriorly to pectoralis magnus and antea spinatus at their insertion; the **anterior scapular** between subscapularis and antea spinatus to the anterior dorsal fossa of the scapula; the **posterior scapular or subscapular** artery, which passes under teres internus, and runs between this muscle and the caput magnum of triceps extensor brachii, to the posterior angle of the scapula; and the **external thoracic artery**, which runs backwards to the lateral part of the thorax, under latissimus dorsi. Inferiorly the **humeral artery** runs from the axillary artery towards the elbow-joint, sending off several constant branches, and finally breaking up into the *ulnar, spiral, and radial branches:* the **ulnar** runs to break up about the point of the ulna, sending an important branch down the limb beneath ulnaris accessorius; the **spiral** twists round the front of the humerus; the **radial** runs along the internal surface of the limb.

I have thus particularised these arteries, since with each of these runs a nerve from the **axillary plexus**, which, being formed by branches from the sixth, seventh, and eighth cervical nerves, and the first dorsal (with a sympathetic branch), passes between the long and short heads of scalenus, and breaks up into the *humeral thoracic, external thoracic, anterior and posterior scapular, ulnar, spiral, and radial nerves.*

By division of the pectorals we have exposed the **lateralis sterni**, running in a downward and backward direction from the posterior border of the lower third of the first rib, to be attached below the inferior extremity of the fourth sterno-costal cartilage to the lateral part of the sternum. This muscle being removed, we have the **intercostales interni** as continued between the sterno-costal cartilages exposed to view; they run in a downward and forward direction, obliquely from the anterior border of one cartilage to the posterior border of the one immediately in front of it. Posteriorly they are covered by the anterior extremity of rectus abdominis, which is anteriorly attached to the external surface of the three or four posterior sternal cartilages of the true ribs.

Having examined these structures, we remove the aponeurotic layer covering **serratus magnus**, and thus expose its inferior attachments. It will be found to arise from the transverse processes of the five posterior cervical vertebræ, and from the external surface of the eight true ribs, having four digitations posteriorly with obliquus abdominis externus; from these points its fibres, forming a large fleshy slab of muscle, converge towards the superior margin of the venter surface of the scapula, into which they

THE ABDOMEN. 219

become inserted between the attachments of rhomboideus and subscapularis, some of the fibres running to the external surface of the cartilage of elongation at its posterior part. This, with the corresponding muscle of the opposite side, forms a sling by which the body is anteriorly supported.

The following are the muscles by which the fore extremity is connected to the trunk:—panniculus carnosus, levator humeri, rhomboideus, trapezius, latissimus dorsi, pectorales magnus, anticus, and transversus, subscapulo-hyoideus, and serratus magnus.

The faschia superficialis abdominis must be now removed with care, for on the proper removal of this depends the successful demonstration of the aponeuroses of the abdominal muscles. If it has become too dry for removal, it is better to raise it with the tendon of obliquus abdominis externus, the fibres of which run downwards and backwards towards the **linea alba**. This is a thick elastic band, formed by the general union of the abdominal muscles at the middle line of the abdomen, extending from the ensiform cartilage to the symphysis pubis, before arriving at which it becomes broader and receives fibres from the several tendons of the abdominal muscles, and a distinct band of tendinous fibres from the antero-inferior spinous process of the ilium, **Poupart's ligament or the crural arch.** Into this ligament (which is really but a portion of obliquus abdominis externus), the posterior fibres of the external oblique are inserted, and between these and those inserted into the linea alba is the **external abdominal ring**, through which passes the spermatic cord, covered by the cremaster muscle. It runs upwards from the external ring posteriorly towards the antero-inferior spinous process of the ilium, the cremaster becoming lost in the lumbar faschia; the remainder of the cord resting on obliquus abdominis internus, at length passes between its fibres and gains the abdominal cavity at the **internal abdominal ring**, the space between the two rings being the **inguinal canal**. This is the passage between the external oblique and the internal oblique muscles; it terminates superiorly at the internal abdominal ring, which is formed by the passage of the prolongation of the peritoneum from the floor of the abdomen to line the scrotum, forming the *tunica vaginalis scroti;* this passes between fibres of the internal oblique muscle. Situated at about the anterior part of the posterior third of the linea alba is the **umbilicus**; this is a slight puckering of the layers forming the floor of the abdomen to close up an opening which, in the fœtus, admitted the passage of the umbilical arteries and the urachus to the fœtal membranes, and the umbilical veins from them; it becomes impervious in extra-uterine life, being no longer needed.

Obliquus abdominis externus is attached to the external surface of the fourteen posterior ribs almost as low down as their cartilages, having anteriorly four fleshy digitations with serratus magnus. From these points its fibres run obliquely downwards and backwards to the under surface of the ensiform cartilage, the linea alba, symphysis pubis, antero-inferior spinous process of the ilium and lumbar faschia. Some of the fibres are arranged to form Poupart's ligament (*vide suprà*) into which others are inserted, while some run to become attached by aponeurosis to the internal surface of the hind limb. Removing this muscle we expose some thin tendinous fibres running obliquely downwards and forwards; these belong to

Obliquus abdominis internus, which seems to be composed of fibres radiating from the antero-inferior spinous process of the ilium and the lumbar faschia and the transverse processes of the three anterior lumbar vertebræ. The most superior of these run to the last rib, and being slightly separated from the rest are sometimes considered as a distinct muscle and termed **retractor costæ**; a few run to the internal surface of the posterior cartilages of the false ribs, most of them running to the linea alba along its whole length, the posterior ones separating to form the internal abdominal ring and the inner wall of the inguinal canal.

Extending for a considerable distance on either side of the linea alba, forming the under part of the wall of the belly, we see a muscle wide centrally and growing narrow anteriorly to become attached to the under surface of the ensiform cartilage and to the external surface of the four posterior sterno-costal cartilages of the true ribs. Posteriorly it grows narrow to become attached to the thick tendon running to the anterior part of symphysis pubis; its inner margin is attached along the whole length of linea alba, and it presents many transverse tendinous markings, by which it is considerably strengthened; this is **rectus abdominis**, its internal surface lies in contact with a few

THE ABDOMEN. 221

scattered tendinous fibres apparently belonging to the internal oblique muscle. Having removed these we see transversalis abdominis, on the external surface of which certain *nerves rendered conspicuous by their whiteness run directly from the intervertebral lumbar and posterior dorsal nerves, in a direction similar to the fibres of this muscle,* which run perpendicularly to the linea alba. On the external surface of this muscle we also see **the arteries which supply the abdominal walls with blood** and their accompanying veins. Superiorly *lumbar branches,* derived directly from the posterior aorta, and the **circumflex artery of the ilium,** which arises from the external iliac artery, runs directly to the antero-inferior spinous process, and then breaks up in supplying the abdominal muscles. Posteriorly the **epigastric** is given off by arteria profunda femoris below symphysis pubis, and passing over the internal surface of the femoral artery and of the inguinal canal, runs forwards under rectus abdominis; it sends off the *external pudic* artery to the sheath in the male, the mammary gland in the female; in the male also the external pudic artery gives off the *anterior dorsal artery of the penis,* one branch of which runs forwards along the superior part of the organ as far as the glans, the other backwards to anastomose with the posterior dorsal artery from the internal pudic.

The *internal thoracic artery* at the ensiform cartilage divides, sending one branch, the *anterior abdominal,* along the floor of the abdomen, the other upwards along the inferior extremities of the ribs, where it anastomoses with the intercostals, and sends twigs to the abdominal walls: all these arteries anastomose by their terminal twigs one with the other.

Transversalis abdominis arises from the internal surface of the twelve posterior ribs, of their cartilages, and by faschial expansion from the transverse processes of the lumbar vertebræ, its fibres run to be inserted perpendicularly into the upper surface of the ensiform cartilage and the linea alba. Anteriorly inside the ribs its attachments digitate with those of the diaphragm. Posteriorly its fibres grow gradually more and more sparse, none being found at the internal abdominal ring. By removing this muscle we expose the **peritoneum** or serous lining membrane of the abdomen covered in places by white fibres which serve to strengthen it. It is transparent and through it we see the cæcum and colon, with their longitudinal muscular bands and resulting puckerings.

The **abdominal cavity** is bounded anteriorly by the diaphragm, inferiorly and laterally by the abdominal muscles, superiorly by the psoæ muscles attached underneath the lumbar vertebræ, posteriorly by the pelvis with its viscera. Its anterior surface is concave, in consequence of the concavity of the diaphragm against which the liver rests on the right side, the stomach and the spleen mainly on the left.

On laying open the cavity from below at the linea alba we expose the various viscera all covered by **peritoneum**, so that none of them can be touched except after removal of their investing serous layer; it likewise forms several double reflections around the viscera, which serve to support the organs and which receive special names. After covering the floor and the sides of the belly it runs upwards posteriorly to be reflected to form the **ligaments of the bladder**; from the floor of the belly along the central line a double fold runs to the fundus or anterior portion of the bladder, being the *suspensory ligament*. From the sides *broad ligaments* run horizontally to the fundus, and on the free margin of each between the folds of peritoneum, is situated the *round ligament* of the bladder, which in the fœtus is the umbilical artery; it arises from the artery of the bulb, and in the adult becomes impervious. The broad ligaments separate to enclose the bladder, and the layer forming the upper surface of these ligaments and covering the upper surface of the bladder is posteriorly inflected upwards on to the under surface of the commencement portion of the rectum and the single colon, on either side of which the membrane becomes firmly attached, superiorly forming by the union of the two layers the **meso-rectum**, by which this portion of the intestine is supported against the spine, and which extends almost as far forwards as the stomach. On arriving at the lumbar region it gives off a pouch on the right side in which are enclosed the double colon and the cæcum and which forms the **meso-colon**, and the **meso-cæcum**, and another on the left in which the small intestines lie, which is termed the **mesentery**; from this posteriorly a smaller fold extends on the right side around the duodenum which it connects to the porta of the liver, and from the anterior extremity of the bowel is continued one layer over the antero-inferior surface of the stomach. After tracing the peritoneum thus far from behind we may commence to examine it anteriorly. Passing forwards from the umbilicus on the floor of the abdomen it forms centrally a longitudinal fold which runs from behind forwards, extending farther back in young subjects than in old; this is one of the **ligaments of the liver**, *falciform or suspensory ligament*. It runs on to the posterior surface of the diaphragm and extends as far as foramen dextrum; on its free margin, enclosed between the two layers, is an impervious fibrous cord, *round ligament*, which is the degenerated umbilical vein of the fœtus by which the blood is carried from the umbilicus to the liver. It is this ligament which is sometimes inquired after as being "the only ligament of the liver not formed by peritoneum." The layers of serous membrane forming the falciform ligament separate superiorly and in surrounding the posterior vena cava in its course from the liver to the foramen dextrum form the *coronary ligament*. The peritoneum then forms the *right lateral ligament*, running from the superior margin of the right lobe of the liver; this ligament is broad and wide, and on its posterior surface we may see *a small special ligament passing to the lobulus Spigelii;* its posterior layer is reflected from the roof of the abdominal cavity at the right kidney. A corresponding ligament, but longer and narrower, runs from the posterior surface of the left side of the diaphragm to the left extremity of the superior margin of the liver, which is the *left lateral ligament.* After investing the liver the serous membrane forms the *gastric hepatic omentum* which runs from the transverse fissure of the liver to the lesser curvature of the stomach and to the attachment of the duodenum to the spine; between its layers run the hepatic and pancreatic ducts. The stomach will be found to be enclosed in a sac, which, in consequence of the fatty matter between its layers arranged around its vessels, is apparently reticulated. This is the *omentum major or gastro-colic omentum* (*caul*), which forms a peculiar cavity connected with the general peritoneal cavity by a channel, **foramen of Winslow**, which is situated below the spine against the cœliac axis. The reflections of peritoneum from the stomach to the neighbouring viscera are termed **omenta**: we have noticed the gastro-hepatic, and gastro-colic omenta; from the left extremity of the greater curvature the *gastro-splenic omentum* passes to the spleen, which is suspended between its two folds; and to the extreme left part of the greater curvature the *gastric-phrenic omentum* is reflected from the diaphragm, it terminates in forming the *cardiac ligament* surrounding the terminating portion of the œsophagus.

In the male at the internal abdominal rings on either side at the posterior part of the cavity the peritoneum is reflected through the inguinal canal into the scrotum, forming *tunica vaginalis scroti*, which by a serous frænum is connected with *tunica vaginalis testis*, a reflection from the roof of the cavity. There is also a *slight fold connecting the two terminal portions of the vasa deferentia together, between which lies the third vesicula seminalis*. In the female from the upper surface of the bladder it passes to the anterior portion of the vagina and invests the uterus, from the horns of which it passes to the roof of the cavity, forming the *broad ligaments*, which anteriorly contain the ovaries and fallopian tubes, and which have the *round ligaments of the uterus and those of the ovary* between their double layers. Between the two horns the serous membrane is reflected backwards over the upper surface of the body of the uterus to the vagina, from which it passes upwards to the under surface of the anterior part of the rectum.

By incision through the terminal portion of the œsophagus, through the rectum, and through their serous connections to the neighbouring organs, especially up against the spine, the intestines and stomach may be removed from the abdominal cavity. In the process we expose the **pancreas**, a body resembling a salivary gland in structure, appearance, and almost in function. It assumes the form of a number of greyish white lobuli placed around the anterior mesenteric artery, moulded in between the various structures in this neighbourhood. It is mainly attached to the double colon, from which it requires careful separation, and will be found for the most part on the right side of the spine, which portion is termed its *head*, and around which the duodenum curves in its upward course. Its various lobules give off ducts which unite and reunite to form finally the **pancreatic duct**, which passes into the gastro-hepatic omentum, where it unites with the biliary (or hepatic) duct and terminates at the duodenum. It is supplied with blood from the cœliac and anterior mesenteric arteries ; its blood passes into vena portæ.

The **liver** is another important accessory gland of digestion. It is that chocolate coloured body found in the anterior part of the abdomen inclined to the right side lying in contact with the diaphragm anteriorly, while the duodenum courses from below upwards over its posterior surface. It is thickest in the centre, and grows gradually thinner towards its outer circumferent margin. It is divided by deep fissures into lobes, of which there are three principal and several accessory. The **right lobe** is the largest, and is connected to the left by the middle. Its anterior surface is smooth, invested by peritoneum; its superior margin presents a depression, *fossa renalis*, on which the right kidney rests; from this part the right lateral ligament runs, and superficially through the superoanterior part of the right lobe, the posterior vena cava passes from the right side of the spine to foramen dextrum. The posteror surface of the right lobe presents superiorly a peculiar triangular accessory lobe, the *lobus Spigelii vel caudatus*. The external and inferior margins are free while the inner is in connection with the **middle lobe**, which anteriorly receives the falciform and round ligaments. Posteriorly, it presents the *porta or transverse fissure*, which extends to the right and left lobes, and to which the arteries resulting from division of the hepatic branch of the cœliac axis pass, and also the vena portæ and gastrohepatic omentum, between the folds of which the nerves and lymphatics run, and also the **hepatic duct,** which meets the pancreatic duct, and the vessel thus formed opens into the duodenum. (There is no gall-bladder in the horse, hence no cystic duct, as in other animals.) The inferior margin of the middle lobe is divided by numerous fissures into several almost equal parts, between two of which the round ligament passes in terminating. This is the *lobus scissatus*. The left margin of the middle lobe blends with the right of the **left lobe,** to the superior margin of which the left lateral ligament is attached. Its other margins are regular, and its surfaces smooth. On carefully examining the surface of the liver we may see that it is mapped out into minute polygonal divisions or *lobuli*, which are found to be shaped like an ivy leaf and consists of a number of hexagonal epithelial "bile cells" recognizable by their yellow colour, due to granular contents. These *acini* or *lobules* are embedded in the special stroma of the gland, **Glisson's capsule**, which has a thin layer investing the surface of the organ and becomes well marked at the transverse fissure. The **circulation through the liver** is complex. **Vena portæ** receives the blood from the anterior and posterior mesenteric, pancreatic, splenic, and (lastly) gastric veins; it passes through a ring formed by the pancreas and direct to the porta of the liver, during its whole course inferiorly placed to the posterior vena cava. It enters the left extremity of the transverse fissure and sends a branch to each lobe, which breaks up into *vaginal branches* found in Glisson's capsule, which divide and subdivide, at length forming the *interlobular veins*, which ramify on the surface of the lobule sending branches inwards; these radiate towards the centre of the lobule, where they terminate in the *intralobular vein*, which at the base of the lobule opens into a *sublobular vein*, in relation to which the acinus is placed as a sessile leaf is to a branch. By union and reunion the sublobular veins at length form the **hepatic veins**, which terminate by numerous orifices at the posterior vena cava, as may be exposed by slitting up in its interhepatic portion. The **hepatic artery** is a branch of the cœliac axis; gives off a branch to the duodenum, which gives off the right gastric artery, and terminates by breaking up before arriving at the porta. Its mode of termination with regard to the hepatic and portal veins is not yet known, its subdivisions have been traced into the substance of the lobuli. The liver is supplied with nerve force by the sympathetic and pneumogastrics.

The abdominal portion of the alimentary tract consists of the stomach and the major portion of the intestinal canal. It presents three coats, peritoneal, muscular, and mucous. The peritoneal coat of the stomach is continued on each side as the omenta; that of the intestines has been described. It is named differently in different parts. Thus the duodenum is supported by a continuation of the gastro-hepatic omentum, the remainder of the small intestines by the mesentery, and the several parts of the large intestine by the mesocæcum, mesocolon, and mesorectum respectively. The **mesentery** is arranged in a spiral manner around the anterior mesenteric artery. In its free margin, between its folds lie the jejunum and the ileum. At first short, it gradually increases in length until it gains its maximum, when it again grows short. Between its folds are arteries, veins, lacteals (or lymphatics), and nerves. The arteries are *mesenteric branches,* from thirteen to eighteen in number, which pass from a common centre and radiate, being connected, forming arches at the attached border of the small intestine, from which smaller vessels are given off, which pass to break up in supplying the coats of the bowel. The *veins,* arranged in like manner, terminate in forming the anterior and posterior mesenteric veins, which form the commencement portion of the vena portæ. The *nerves* are some of those branches of the sympathatic which, in radiating from the semilunar ganglion, form the solar plexus. The *lacteals* are lymphatic vessels, which being filled with an opaque milky fluid, chyle, the product of digestion, receive a name distinct from those in other parts of the body. Passing from the intestinal villi, these minute vessels proceed upwards towards the spine and near the anterior mesenteric artery pass through the *mesenteric glands,* numerous greyish nodular bodies, by which the lacteal fluid is rendered cellular and coagulable; they differ in no point in structure from other lymph-glands of the body. The termination of the lacteals is in the **receptaculum chyli**, a dilated vessel which receives lymph from the hinder parts of the body and the chyle, and from the anterior part of which the thoracic duct runs forwards and serves to convey the fluid into the venous circulation. This cavity has very thin transparent walls, and is situated between the anterior mesenteric artery and the cœliac axis, in which situation also we may find the **semilunar ganglion**, the largest ganglion in the body, to which the splanchnic and part of the pneumogastric nerves pass, and from which plexus run with the branches of the cœliac axis, anterior and posterior mesenteric, and renal arteries receiving their names after the arteries they surround. Those in the mesentery form the **solar plexus**. Branches run backwards to the **hypogastric plexus**.

The peritoneal coat of the double colon is reflected from the under surface of the loins and covers at once both the diverging and returning portions, so that they are contained in a common serous covering terminating in a blind extremity at the sigmoid flexure. That portion between the two divisions of the intestine, which is most marked at the sigmoid flexure, is termed the mesocolon. It presents the figure of a racket bat, the broad portion of which is at the sigmoid flexure. The diverging portion of the double colon is also attached to the cæcum by a double fold of peritoneum, the mesocæcum, while the single colon floats at the free margin of a double peritoneal fold, which, after being continued for a slight distance posteriorly around the anterior part of the rectum (the mesorectum), is reflected laterally on to the walls of the abdomen in forming the anterior boundary of the pelvis. In the large, as in the small intestine, we find the arteries forming loops between the peritoneal supporting layers from which branches run to the attached margin of the intestine, where they break up in supplying mainly the muscular and the mucous coats. The cæcum is supplied by two branches of the anterior mesenteric artery. They run towards its apex and anastomose just before reaching it; they send off numerous branches. The double colon is supplied by two branches from the *anterior mesenteric*, which run along the attached margins of the diverging and returning portions respectively, and anastomose on converging at the sigmoid flexure. A branch from the anterior mesenteric also passes to the anterior part of the single colon, and posteriorly joins the first branch of the **posterior mesenteric**, which breaks up in the mesorectum, forming arched vessels which supply the single colon and the anterior part of the rectum, anastomosing posteriorly with the hæmorrhoidal branches of the lateral sacral, by which, in conjunction with branches of the artery of the bulb, the posterior part of the rectum and the anus are supplied. The *veins of the large intestine* converge to form the anterior and posterior mesenteric veins, which form the commencement of vena portæ, up against the anterior mesenteric artery. The *nerve supply of the large intestine* is conveyed by branches from the semilunar and *posterior mesenteric ganglia*. The latter is found in contact with the artery of the same name, and sends fibres in a backward direction, which combine to form an intricate arrangement of nerve-fibres of the walls of the rectum in the pelvis, to which the name *hypogastric plexus* is given. Numerous lymphatic or lacteal glands may be seen in apposition with the attached portions of the large intestine along its whole length, but predominating in the anterior part. The vessels from thence pass towards the spine to empty themselves into the receptaculum chyli.

The **stomach**, somewhat small in the horse as compared with other animals, is found mainly in the left hypochondriac region of the abdomen. It however extends into the epigastric, and sometimes into the right hypochondrium under extreme distension. It is likened to the bag of a bagpipe, and from this it may be seen that it is a somewhat elongated body, curved upon itself, having two extremities. To one of these extremities, that placed on the left side and most superiorly, in contact with the posterior surface of the diaphragm, the œsophagus passes after emerging through foramen sinistrum. It is surrounded by a fold of peritoneum, which is designated the *cardiac ligament*, which posteriorly expands over the stomach, while to the left it forms an omentum which serves to connect the stomach to the diaphragm, and is termed the *gastrophrenic omentum*. This extremity of the stomach is termed the **left or cardiac**, while from the **right or pyloric extremity** the small intestine commences. That margin of the organ situated between these two extremities, and which is most superiorly placed, passing in an oblique direction from left to right and from above downwards, is termed the **lesser curvature**, and from it the gastrohepatic omentum runs to the porta of the liver, and on the right side extends to the spine, having the duodenum between its layers at the free margin, to which the duct formed by combination of the excretory vessels of the pancreas and liver runs, almost directly from the liver to the bowel, into which it opens about four inches from the pylorus, on the free margin, having wound round the intestine for that purpose. The two extremities of the stomach being attached to the upper wall of the abdomen, the **greater curvature** of the stomach is situated below and behind the lesser in such a manner that the organ being obliquely placed from above downwards from left to right, as well as from before to behind, presents a *postero-superior* and an *antero-inferior surface*, both of which are covered by layers of peritoneum, which at the greater curvature form omenta. Thus, to the left extremity is attached the spleen by the gastro-splenic omentum, while to the greater part of the greater curvature the gastro-colic omentum is attached, and, by running in a backward direction, and then again forwards, and forming a covering for part of the double colon, produces a peculiar cavity of serous membrane, connected to the main cavity by a channel up against the anterior mesenteric artery, which has received the name of **foramen of Winslow**. In the gastro-colic or or omentum major, we sometimes find much fat. By butchers it is termed the " *caul*."

The stomach of the horse presents a rounded cul-de-sac at its left extremity, more marked than in most other animals. It has a *muscular and a mucous coat* in addition to the peritoneal already described. The **muscular coat**, centrally placed between the serous and mucous layers, presents *three layers of fibres : longitudinal*, most marked along the lesser curvature, continuous with those of the œsophagus and duodenum at their extremities ; *oblique*, which predominate at the left extremity ; and *circular*, continuous with those of the œsophagus and duodenum, extending over the whole surface of the viscus, and which centrally by their contraction seem sometimes to practically divide it into two compartments.

The **mucous layer** is divided into two parts, *cuticular and villous*. These two portions present a sharp line of demarcation, which is sigmoid. The former occupies about the left third of the inner surface of the stomach, and is analogous with the mucous membrane of the three first compartments of the stomach of the ruminant, for it is directly continuous with that of the œsophagus, being whiter and more firm than that forming the villous portion, which is reddish in colour, slimy in nature, and occupies the left two thirds of the stomach, being continuous through the pyloric opening with that of the duodenum. Upon examination under the microcsope we may distinguish in this portion glandular follicles or **peptic glands**, the deeper portions of which are lined by *spheroidal* epithelium, and which secrete the gastric juice, in addition to those mucous follicles lined by *tessellated* epithelium, which secrete mucus, and are apparent also in the cuticular portion. To the cuticular portion we sometimes find bots (œstrus equi) adhering, rarely to the villous. The left extremity of the stomach comes in apposition with the pancreas, diaphragm, left kidney, and left pre-renal capsule. *All the branches of the cœliac axis supply this organ with blood.* The gastric (centrally situated) runs direct to the centre of the lesser curvature, where it breaks up in forming the *superior* and *inferior gastric arteries*, which run to the surfaces of the same name. The splenic, inclining to the left, passes between the layers of the gastro-splenic omentum, sending branches outwards to the spleen, inwards to the stomach, and, after gaining the apex of the spleen, it continues along the greater curvature as the *left gastric artery*. The hepatic, running to the right side to gain the porta of the liver, sends a branch to the duodenum, which, on gaining the intestine, sends a branch backwards to anastomose with the first mesenteric branch of the anterior mesenteric trunk, and one forwards to be continued beyond the pylorus, over the right extremity of the greater curvature, forming the *right gastric artery*. All these vessels give off branches freely, which anastomose, and from which the blood is returned by the **gastric vein**, which is the last to gain vena portæ before it enters the liver. The *nerves of the stomach* are derived from the splanchnic and solar nerves; while the left pneumogastric mainly expends itself on the upper surface, the right runs to the inferior, but both these nerves are intimately connected with the semilunar ganglion. For convenience of reference the **cavity of the belly** is by anatomists artificially divided into nine parts. From either side of the ensiform cartilage in a direct line backwards to the brim of the pelvis, and from the antero-inferior spinous process of one ilium to the other, and from the posterior margin of the last rib to that of the other, lines are drawn, and in this way the cavity is found to present anteriorly the *epigastric portion*, with the *right and left hypochondriacs* on either side; centrally, the *umbilical portion*, with the *right and left lumbars*; posteriorly, the *hypogastric* and the *right and left iliacs*.

The **intestines** present three coats—*serous, muscular,* and *mucous*. The **serous** coat has already been noticed as part of the visceral peritoneum. It is the most external of the coats, and extends from the stomach to the anterior part of the rectum, thus leaving the posterior part of that bowel with only two coats. It is also deficient where two portions of intestine lie in close apposition *permanently*, or where organs, as the pancreas, come into intimate contact with the intestinal walls. The **muscular** coat consists of two orders of fibres of the white, unstriated, or involuntary class. The *internal or circular*, which run round the intestinal tube, are found in all parts evenly diffused over the surface of the intestine, being anteriorly in direct connection with the circular fibres of the stomach at the pylorus. Posteriorly collected into a mass, which forms the internal portion of the sphincter ani. The *external or longitudinal order of fibres* is evenly diffused over the surface of the small intestine and of the rectum, while in the cæcum and colon these fibres are collected into thick bands, which run longitudinally to the course of the bowel, and are more or less numerous in different parts. Being shorter than the bowel itself, and at the same time intimately attached to its walls, they serve to "pucker" up its surface into a number of cavities, which give it a peculiar appearance, whilst by division of the bands it may be reduced to a tube of uniform diameter and of much greater length. On exposing the bowels of an animal shortly after death we may observe the *vermicular or peristaltic action of the intestines*, that slow consecutive contraction of their fibres commencing anteriorly and extending backwards, whereby the ingesta is slowly propelled from the pylorus to the anus. The **mucous** coat of the bowels is of a reddish-grey colour and slimy nature, extends from the pylorus (where it is continuous with the villous portion of the gastric mucous membrane) to the anus, where it gradually blends with the common integument. It consists of corium and basement membrane, on the free surface of which is *columnar* epithelium, and throughout presents depressions of this mucous membrane termed *follicles of Lieberkühn*, which secrete mucus, and prominences of the same, *villi or papillæ*, from the centre of which a lacteal commences, either by a plexus, loop, or a blind pouch, for these are the portions of the intestine from which the lacteals arise, and, though most numerous in the small intestine, they may be found also in the large. Throughout the intestinal tract also we find embedded in the mucous membrane small bodies called *solitary glands*, which consist apparently of a number of cells bound together by somewhat dense areolar tissue, and having blood-vessels running through them. They present no ducts, and hence are considered to be analogous in nature and function with the Malpighian corpuscles of the spleen, though some authorities assert that by bursting they discharge their contents into the intestine, and thus give the fæces its peculiar odour. The intestines of the horse are about ninety feet in length, and as the posterior eighteen feet are much larger in calibre than the rest, it is divided into *the large and the small intestine*. The **small intestine** commences from the pyloric extremity of the stomach, and just after its commencement becomes very large. It is here termed the **duodenum**, and takes an upward direction over the posterior concave surface of the liver, to the porta of which it is attached by a continuation of the gastro-hepatic omentum, between the two layers of which the duct common to the liver and the pancreas runs to the attached margin, and then, after passing beneath the peritoneal coat, opens on the free margin through a perforated papilla into the intestinal canal, about four inches from the pylorus. Passing in an upward direction, the duodenum takes a curve around the head or right extremity of the pancreas (termed its *horse-shoe curve*), and then, passing underneath the spine *behind* the anterior mesenteric artery, terminates. The rest of the small intestine occupies no regular position in the abdomen, for, being attached in the free margin of the mesentery, it hangs loose, but terminates in the right lumbar region, where it opens into the *large intestine*, the muscular coat being here doubled in such a manner as to form the **ileo-cæcal valve**, which prevents food from passing from the large into the small intestines under ordinary circumstances. The anterior two fifths of this portion of the small intestine is termed the *jejunum*, and the posterior three fifths the *ileum*. This division is purely artificial, and anatomists now prefer to describe the *fixed or duodenal*, and the *free portions of the small intestine*. The mucous coat of the duodenum presents, in addition to papillæ, follicles of Lieberkühn, and solitary glands (extremely rare here), certain small racemose glands, in every respect resembling salivary glands, even, it is supposed, in the nature of their secretion, of which, however, nothing definite is known, these are named *Brunner's glands*, after their discoverer. On the free margin of the ileum and of the posterior part of the jejunum may be seen elongated patches, apparently of pigmentary matter. These are *Peyer's patches*, and when examined microscopically are found to be aggregations of solitary glands surrounded by a row of follicles of Lieberkühn termed the *corona tubulorum*, while villi may be seen covering the space enclosed by the corona.

The **large intestine** consists of three portions: *cæcum, colon, and rectum.* The ileum terminates at a peculiarly arched portion of intestine, the *cæcum caput coli,* at its lesser curvature, anteriorly is the commencement of the double colon. The **cæcum,** a plicated portion of intestine, presents a base situated in the right lumbar and right iliac regions, from which it extends in a direction forwards and to the left, through the umbilical region along the floor of the abdomen, to terminate in the *apex* in the left hypochondriac region. Its plicæ are due to the collection of its longitudinal muscular fibres into three bands at the apex, while subsequently a fourth appears. This viscus is capable of containing about four gallons of fluid, and is supposed to be the destination of the huge draughts of water sometimes consumed by horses, which rushes directly through the stomach hither; and since it is apt to carry with it any undigested matters, we learn from this the rule found valuable in practice—"to water before feeding."

The **colon** is divided into two portions—*double and single;* for after proceeding a certain distance the bowel bends backwards and returns on the same course, becoming attached to the diverging portion by the meso-colon, and in some parts closely in contact with it. The posterior part of the colon, however, runs a single course, being situated at the free margin of the mesorectum, and like the free portion of the small intestine having no regular situation. It takes the following course:—Commencing at the right lumbar region with two muscular bands, it runs in a forward direction from the cæcum caput coli into the right hypochondriac region, where it obtains a third, and subsequently a fourth longitudinal band; passing over to the left, above the ensiform cartilage, it gains the left hypochondrium, and, in running from this backwards, grows gradually smaller in calibre as it passes through the left lumbar into the left iliac region. Its bands also decrease in number, so that when it makes its sigmoid flexure in the left iliac region it has but one, and that is on the attached margin. The bowel is here very small, though not so small as at its commencement, and in passing forwards through the left lumbar and left hypochondriac regions, superiorly placed to the diverging portion, it increases in size and the number of its bands until it again reaches the epigastric region, when it inclines backwards, it attains its maximum calibre, and has three muscular bands. After passing into the umbilical region it terminates in the single colon, which has two muscular bands, hangs freely in the free margin of the mesorectum, being about three feet in length, until at the central line posteriorly it terminates in forming the **rectum**, which runs directly from the posterior part of the hypogastric region to the anus. Its muscular wall is very thick with the fibres well marked; superiorly it is in contact with the under surface of the sacrum and the structures about it; while its inferior surface, in the male, rests upon the upper surface of the bladder, prostate, &c., and in the female upon the vagina. In contact with its walls laterally is the hypogastric plexus. Its muscular coat terminates posteriorly in forming the inner portion of **sphincter ani**, of which the external portion is formed of dark-red fibres, *supplied with nerve-force by the posterior part of the spinal cord*. To this muscle superiorly is attached **levator ani**, running from the under part of the anterior coccygeal bones. Laterally, **retractor ani**, from the inner surface of the sacro-sciatic ligament, as far forward as the ileum; inferiorly, **depressor ani**, which, in the female, blends with sphincter vaginæ; in the male with retractor penis. The mucous membrane of the rectum is continuous posteriorly with the thin, hairless skin, which is around the anus "like the mouth of a draw-purse," and which has many sebaceous glands. Some fat also enters into the formation of the anus. The glands and villi of the mucous membrane gradually decrease in number from the cæcum to the anus.

The **spleen** is a peculiar sickle-shaped organ, attached to the left extremity of the greater curvature of the stomach by the gastro-splenic omentum. Of a bluish colour, and about three pounds in weight; it presents *a base and an apex, a greater and a lesser curvature, and two surfaces.* The *base* is attached up against the left kidney, and prerenal capsule, and from this the organ extends in a downward and backward direction and towards the right side, gradually decreasing in size. The *lesser curvature, hilum,* is that situated nearest the stomach along which the vessels of the organ run; from this the organ decreases in thickness towards the *greater or free curvature*, around which the serous membrane is reflected joining the layers on the surfaces; the curvatures meet at the *apex*. The special *capsule of the spleen* consists mainly of yellow elastic tissue, whereby it is endowed with great capacity of distension, for it is an erectile organ. From this outer covering *trabeculæ or septa* project inwards in all directions, thus dividing the organ into spaces which contain a peculiar reddish matter, *spleen pulp*, aptly compared to raspberry jam. On examination under the microscope this is found to consist essentially of large cells, each containing one or more red cells, in a more or less advanced state of disintegration, for it is supposed that the spleen is the organ whereby excess of red corpuscles in the blood is prevented. The spaces between the trabeculæ also contain vessels. The **arteries**, derived from the splenic branch of the cœliac axis, *terminate in venous sinues*, and are also remarkable for presenting at various intervals, embedded in their fibro-areolar coat, small bodies analogous in structure to the solitary glands of the intestines. These are **Malpighian corpuscles**, the use of which is not known, and which consist of aggregations of cells embedded in a fibrous stroma. The *veins of the spleen* commence from the venous sinuses, *they have few if any valves*, and pour their contents into the vena portæ. The nerves of the spleen are derived from the semilunar ganglion. The **prerenal capsule** is another organ the use of which is undetermined. There is one on either side of the body, situated between the kidney and the spine, and presenting two surfaces. It derives its blood by a special arterial branch, either from the posterior aorta or from the renal artery. When cut into it presents an external dark, or *cortical portion*, the use of which is unknown; and a central lighter *medullary part*, supposed to be a nerve centre. It is included among the ill-understood ductless or vascular class of glands.

Having removed the alimentary viscera from the belly we are enabled to examine the structures situated in the upper part of the cavity. On the central line we see important vessels and nerves. The **posterior aorta**, after passing through hiatus aorticus of the diaphragm in company with the thoracic duct and vena azygos, runs backwards, inclined to the left side of the spine. In its course it gives off important vessels. **Cœliac axis**, a single trunk, is but a short vessel, for it soon breaks up into three branches, gastric, splenic, and hepatic. The **gastric** runs direct from its commencement to the central part of the lesser curvature of the stomach, where it breaks up in forming the *superior* and *inferior gastric arteries*. The **splenic** inclining to the left side and gaining the base of the spleen runs along its lesser curvature between the two folds of peritoneum which constitute gastro-splenic omentum. During its course it gives off large branches to the spleen, smaller branches to the stomach, and at the apex of the spleen is continued along the greater curvature of the stomach as the *left gastric artery*. The **hepatic artery**, which inclines to the right side, has already been described.

Anterior mesenteric artery (in the ass extremely liable to aneurism from the presence of parasites of the nematode class—strongylus armatus) arises from the under part of the posterior aorta. It gives off a few branches to the pancreas which completely surrounds it. And then the branches : *mesenteric*, thirteen to eighteen in number, which radiate through the mesentery to the small intestine ; the first anastomoses with the duodenal branch of the hepatic, the last with the *cæcal branch*, which in its turn communicates with the *branches to the double colon*, and they with that *to the single colon*, which joins the first branch of the **posterior mesenteric artery**, a vessel supplying the single colon and rectum in a manner already described ; it arises from the under surface of the posterior aorta just prior to its termination.

Renal arteries arise from the lateral part of the posterior aorta just behind the anterior mesenteric trunk. They are large vessels, given off at right angles. They are also termed the **emulgent arteries**, and before arriving at the kidneys break up, *sometimes sending a branch to the prerenal capsule* (which sometimes arises from the posterior aorta). Most of their branches enter the kidney at a particular notch, in its substance termed the *hilum*, some run to other parts of the organ. The inclination of the posterior aorta to the left side of the spine renders the *right* renal artery the longest, for it has to pass under the posterior vena cava. *It is also given off more anteriorly than the left.*

The **spermatic artery** is given off from the posterior aorta, between the commencement of the renal and the posterior mesenteric vessels. It is remarkable for its length in proportion to its size, and for the convoluted manner in which it runs through the spermatic cord. It runs in a downward and backward direction to the internal abdominal ring, where it gains the spermatic cord, in the anterior part of which it passes much curled upon itself. On getting near the testicle it sends off small long branches to the commencement portion of the vas deferens, and on gaining the antero-superior part of the organ divides into two parts, one of which runs in a backward direction, the other downwards until it gains the inferior margin, when it also curves backwards. Both of these divisions send branches into the substance of the organ which help to form the *tunica vasculosa*. Sometimes the **artery of the cord** is given off from the posterior aorta (often from the external iliac). It runs to the posterior part of the cord, supplying the cremaster and vas deferens. These vessels are arranged as above described in the male, but in the female the **ovarian artery** takes the place of the spermatic, being much shorter and smaller in every respect. These conditions, however, are reversed in the **uterine artery**, which much exceeds its analogue the artery of the cord in size, especially during pregnancy. It runs downwards and backwards towards the pelvis between the two folds of peritoneum which constitute the broad ligaments of the uterus.

From its upper surface the abdominal portion of the posterior aorta gives off the **lumbar arteries**, which passing upwards, give branches to the psoæ muscles, others to the muscles on the loins, the medullary arteries to the lumbar vertebræ, and the arteries to the lumbar portion of the spinal cord. Then the posterior aorta gives off the **external iliac arteries** in an outward direction, whereby it is much reduced in calibre, and under the second sacral bone it terminates in forming the **internal iliacs**, and sometimes the **middle sacral**, the former running downwards and outwards, the latter directly backwards. On the right side of the posterior aorta runs the **posterior vena cava**; it commences beneath the anterior part of the under surface of the sacrum, resulting from the convergence of the common iliac veins, and in passing forwards, receives some veins corresponding to arteries given off by its companion vessel, *spermatic vein, posterior lumbar veins* (*anterior lumbar veins combine to form vena azygos*) *and renal veins*. Of the **renal veins**, the left is the longest, but the right anteriorly placed, and shortly after receiving this vessel the post vena cava passes through its special channel in the upper part of the right lobe of the liver. While passing through this it presents the openings, numerous and varying in size, of the *hepatic veins*, and on emerging, is surrounded by the coronary ligament in its short course to foramen dextrum, where it receives the *phrenic venous sinuses*, and then runs straight through the chest, enveloped in a special fold of pleura to the base of the heart.

On removing the peritoneum from the lateral parts of the upper surface of the abdomen, we find that it simply *covers* the under surface of the kidneys, from which it is separated by a quantity of fat, unusually firm in consequence of the predominance of stearin in its composition. This is vulgarly named "**suet**," and by some anatomists has been termed the *tunica adiposa renalis*.

The **kidneys** are the organs which serve to separate urine from the blood. They are two purple-coloured organs situated up against the crura of the diaphragm. The **right** is most anteriorly placed, and differs from its fellow in being heart-shaped, and resting against the diaphragm, and in a special depression on the upper part of the right lobe of the liver termed the *fossa renalis*. The **left** kidney is elongated from behind forwards, being shaped like a "kidney-bean." It receives the shortest artery and the longest vein, and lies in contact with the base of the spleen and with the left extremity of the stomach. Both kidneys have a prerenal capsule in contact with the anterior part of their *inner border* which presents a *notch or hilum*, to which most of the branches of the renal arteries pass, and from which those of the renal vein together with the ureter commence. The anterior part of the kidney is slightly the largest, the outer margin convex, the inner straight, while the under surface is flatter than the upper. The organ is supported by a special stroma (*renal capsule*) which is considerably more abundant in the substance of the organ, investing its vessels, and connecting together its component structures. A horizontal incision displays centrally a cavity lined with mucous membrane, **pelvis**. Externally a reddish-mottled smooth surface, *cortical portion*, and between the two, the *medullary portion*, which is apparent to the naked eye as consisting of tubular fibres, radiating from the central cavity to the cortical portion. By microscopic examination we may find that these tubes, *tubuli uriniferi*, are continued into the cortical portion, when they become much convoluted, forming the *tubuli contorti*, and finally terminate in blind dilated extremities, *Malpighian capsules*.

The **branches of the renal arteries** after division and subdivision by traversing the medullary part, and breaking up in their course gain the cortical portion, where as small *afferent vessels* they run to the Malpighian corpuscles of which they push the mucous membrane before them, causing it to bulge towards the commencement of the tubulus uriniferus, for they exert their pressure at a point opposite to this. On entering the depression of the wall of the corpuscles thus produced, the vessel breaks up into smaller branches, which divide and reunite producing the *glomerulus* or *Malpighian tuft*, from which arises the *efferent vessel*, which emerges from the depression at the point where the afferent vessel enters, and then runs to the wall of the contorted portion of the uriniferous tube, *around which it forms a plexus of capillaries*, from which veins arise, and after passing through the medullary portion of the kidney by union and reunion, combine to form the renal vein. The tubuli uriniferi in the medullary portion, in their course from the cortical portion to the pelvis, unite two by two (dichotomously), and as this occurs frequently, the whole of the tubes opening by one perforation into the pelvis may be represented by a pyramid, the base of which is at the cortical portion, the apex at the pelvis, and in the foetal colt the kidney is lobulated, divided into the "*pyramids of Malpighi*," while the same condition obtains permanently in the ox. The pelvis of the kidney of the horse is a fan-shaped cavity lined by *spheroidal* epithelium, and around its convex border a ridge extends on to which the uriniferous tubes open. This ridge terminates laterally in cavities termed *calyces of the pelvis*, while at the hilum the pelvis is continuous with the **ureter**. This is a musculo-membranous canal extending along the roof of the abdomen, downwards, outwards, and backwards to the pelvis, where it terminates at the supero-posterior part of the bladder. It first passes through the muscular coat of that organ, and after running for a short distance between this and the mucous coat, penetrates the latter, and thus distension of the bladder serves more securely to prevent reflux of urine upon the kidneys. It presents a *muscular coat*, with an external layer of longitudinal, and an internal layer of circular fibres, and is lined with *mucous membrane* which presents epithelial cells of the *spheroidal* character. The ureters run between the two layers of the broad ligaments of the bladder, and hence are by some anatomists described as having an external serous coat. The **bladder** is situated in the lower part of the pelvis, and is a pyriform organ, rounded anteriorly, terminating posteriorly in becoming constricted to form the *neck*. The anterior portion is the *fundus*, and the portion between the fundus and the neck is the *body*. Superiorly it is in contact in the female with the vagina and the uterus, with the rectum in the male; while on either side from before backwards run a vesicula seminalis and a vas deferens. The under surface rests upon the upper concave surface of the pubic portion of the ossa innominata. The fundus presents a remarkable puckering, the result of the change which the bladder undergoes after birth, whereby it gradually decreases in proportionate size, and tends to retract from the umbilicus into the pelvis, from which it extends into the abdomen in a degree varying with its plenitude. To the fundus run the *round ligaments of the bladder*, which diverge at an acute angle and run forwards and outwards, enclosed in the double folds of peritoneum forming the *broad ligaments*, towards the umbilicus. The fundus also receives a coating of peritoneum which forms a small cap for it, extending farthest backwards superiorly and presents double folds running in various directions; thus longitudinally forwards the *falciform or suspensory ligament* runs to the umbilicus. Laterally, the *broad ligaments* (with the round ligaments in their free margins) to the walls of the cavity, and superiorly a longitudinal fold, *vesico-rectal fold* in the male, *vesico-vaginal* in the female. The vesico-rectal fold is bounded laterally by the vasa deferentia, and *encloses the third vesicula seminalis between its layers*. The ureters pass to the supero-posterior part of the bladder, and open into it in the peculiar manner already described. The organ presents *two coats—muscular and mucous*. When it is distended, the mucous coat may be seen between the muscular fibres, which seem to be irregularly arranged, but they present a species of oblique arrangement on closer inspection, forming what is termed *detrusor urinæ*, while towards the neck of the bladder they tend to assume the circular form, producing the *sphincter vesicæ*. The mucous coat lines the bladder throughout, and is continuous with that of the ureters, and with that of the urethra; it presents thick and viscid epithelium, which serves to protect the organ from the irritant action of the urine, and much of which is expelled with that fluid.

The **urethra** is a membranous elongated passage, commencing at the neck of the bladder and running in a backward direction as far as the ischial arch, around which it winds, and then runs in a forward direction forming the central part of the penis, to between the symphysis pubis and the umbilicus. It consists of a *mucous membrane*, which extends throughout, and in some parts presents large *mucous crypts or lacunæ*, which also communicates with that lining the bladder, ejaculatory ducts, prostatic, and Cowper's ducts, and with the lining layer of the prepuce. It is divided into three portions, *prostatic, membranous, and spongy*. The **prostatic portion** is the shortest, and is surrounded by the prostate gland. It commences at the neck of the bladder and passes to the membranous portion. Its mucous membrane presents at its upper part two peculiar prominences perforated by two foramina, the openings of the ejaculatory ducts formed by union of the vasa deferentia and of the ducts from the vesiculæ seminales. This prominence is the *veru montanum* or *caput gallinaginis*. On either side also may be seen a row of small papillæ, on to which open the prostatic ducts. The **membranous portion** extends from the prostatic as far backwards as the ischial arch, and on its postero-lateral parts are two rounded bodies about the size of a walnut. These are *Cowper's glands*, and like this portion of the canal they are invested by a portion of muscle, **Wilson's muscle**, the fibres of which run transversely above the canal from one side to the other. Like the prostatic their ducts open upon numerous small papillæ arranged in rows, one on each side of this portion of the urethral canal. The spongy portion is completely invested by the corpus spongiosum of the penis, and anteriorly dilates, forming the *widest portion or fossa navicularis,* prior to terminating in *its narrowest portion, meatus urinarius externus.* A catheter which will pass through here will pass along any portion of the canal, unless obstructed by foreign matter.

The **prostate gland** is a body which embraces superiorly and laterally the neck of the bladder. It is divided into *two lateral, and a connecting isthmus or middle lobe.* This gland is racemose in its nature, and pours its secretion through the above described openings into the urethra. **Cowper's glands** are two in number, and are covered by Wilson's muscle; they resemble the prostate in structure. The function of these glands is but little known. Their secretion, by mixing with the semen, is considered to increase its efficacy. The prostate gland hides from view the terminal portion of the vas deferens, and of the duct of vesicula seminalis and the ejaculatory duct.

The **penis** is a dense elongated organ extending from the posterior part of the ischium in a forward direction under the symphysis of the ossa innominata. It is mainly composed of the **corpus cavernosum.** This is situated at the superior part, and presents a supero-lateral surface, rounded and covered by a plexus of veins, an inferior surface concave forming a groove through which the urethra (spongy) runs, surrounded by *corpus spongiosum.* Anteriorly it presents two sharp points, which are concealed by the *glans penis;* posteriorly, by bifurcation, it produces the *crura,* dense, yellowish, fibrous bands running to become attached to the tuberosities of the ischium, which are surrounded by the **erectores penis muscles,** arising from the tuberosities of the ischium and becoming inserted into the root of the penis. On transverse section this organ will be found to present a tendency to division into two parts. It has externally a thick layer of yellow elastic tissue, from which trabeculæ and bands are sent inwards, which, by their union and reunion, produce spaces, which are occupied by venous sinuses. The most remarkable of these septa projects from the central line at the superior part. It consists of a number of bands running downwards, resembling in arrangements the teeth of a comb, *septum pectiniforme.*

The penis anteriorly receives blood from the *external*

pudic artery, which sends a branch backwards along the upper surface of the organ to meet the *anterior branch of the internal pudic*, which also gives off a branch backwards to gain the *artery of the bulb*. Immediately investing the surface of the urethra, situated in the inferior groove of corpus cavernosum, is **corpus spongiosum**, a thin layer consisting of erectile tissue analogous to that found in the hard palate serving to connect the mucous membrane to the bony palate. Thus it consists of a network of elastic tissue, through which blood-vessels ramify, and which sometimes presents venous sinuses. Posteriorly, against the crura, it forms a small dilatation termed the **bulb**, which is covered by erectores penis. Posteriorly it forms a large prominent portion, **glans penis**, which, under erection, assumes the shape of the extremity of a trumpet. It completely covers the anterior portion of the corpus cavernosum, and around its outer circumferent margin is a row of glands, *corona glandis*, which secrete a sebaceous material. It is covered by a highly sensitive mucous layer, and at its centre, inclined to the lower part, presents a deep depression, into which the anterior extremity of the urethral canal projects, and which secretes a strong-smelling greyish matter, termed smegma preputii. From one infero-lateral part of the corpus cavernosum to the other, transverse, dark-coloured muscular fibres run underneath the urethra and corpus spongiosum, forming the **accelerator urinæ muscle**, which runs from the glans penis as far back as the erectors. Along its under surface may be seen two bands, apparently of white fibrous tissue, which may be traced as far as the lateral part of sphincter ani. These are the **retractores penis**, composed of involuntary muscular fibres.

The penis as above described is inferiorly covered by skin which communicates laterally with that covering the thighs, superiorly with that portion extending towards the anus over the space termed the **perineum**, and anteriorly it spreads outwards and joins the posterior part of the cuticular layer of the scrotum, and in front of this the outer layer of the prepuce. The **prepuce or foreskin** is a circular inversion of the skin inward to form mucous membrane. Anteriorly it is continued forwards to form a *raphé* in the neighbourhood of the umbilicus. The skin in this region is very thin, and in most horses almost hairless. The mucous membrane lining the prepuce is darkened by pigment, and presents many glands, *glandulæ odoriferæ*, which secrete a peculiar unctuous lubricating matter. It is continuous with that of the glans penis. Between it and the skin is a layer of yellow elastic tissue, reflected from the fascia superficialis abdominis, which, forming a sling beneath the anterior part of the penis, is termed the *suspensory ligament*. From the central part of the under surface of symphysis pubis a fibrous ligament extends, composed of two diverging portions, one of which runs to either side of corpus cavernosum ; this is the *triangular ligament of the penis*, between its two parts runs a vein. This organ serves to direct into the vagina of the female the *semen or male fecundating fluid*. Semen is produced by certain organs, *testes or testicles*, which are suspended from the upper part of the abdominal cavity in certain special cavities, **vaginal cavities**, situated at the posterior lateral and inferior part of the belly. These cavities are covered by layers continuous with those of the floor of the abdomen which form the

Scrotum, *or "purse,"* situated just behind the sheath, which presents externally a layer of thin skin, along the central line of which running from before backwards is an irregular ridge or *raphé*. This covers the **dartos muscle**, composed of unstriated muscular fibres, continuous with fascia superficialis abdominis, which throws across the cavity a septum (septum scroti), whereby the right scrotal cavity is separated from the left. It is contraction of this which causes corrugation of the scrotum under the influence of cold, by means of the *intercollumnar fascia* (derived from obliquus abdominis externus), its inner surface is attached to the *cremasteric fascia*.

The **cremaster muscle** is but a portion of the internal oblique of the abdomen. Arising from the psoas fascia, or from the antero-inferior spinous process of the ilium, it passes through the inguinal canal and the external abdominal ring, externally situated to the spermatic cord, and on approaching the scrotum its fibres form a species of sling serving to support the testis. To this portion the name *cremasteric fascia* is applied, and its internal surface is connected by the *infundibuliform fascia* (occasionally connected with transversalis abdominis) to the external surface of *tunica vaginalis scroti or propria*, a continuation of the serous membrane covering the floor of the belly through the inguinal canal. The external surface of this membrane is smooth, covered with *tessellated* epithelium, which secretes a serous fluid to enable the surface of *tunica vaginalis testis or reflexa* to glide on it freely. These two serous layers are posteriorly connected by a *frænum*, which extends up as far as the loins whence the testes originally descended. In the fœtal colt, as in adult birds, the testis occupies a position against the kidney. To its under part is attached a white fibrous cord, *gubernaculum testis*, surrounded by serous membranes, running to the internal abdominal ring. In the process of development this guides the testis towards the inguinal canal, and then becoming everted in passing into the scrotum, its serous covering becomes the tunica vaginalis scroti. The testis may generally be found in the scrotum at birth; it subsequently, at a period varying according to the breed and treatment of the colt, passes again into the inguinal canal, where it remains for some time, sometimes permanently, and again descends into the scrotum. Sometimes, however, it is drawn up into the abdomen, and on account of increase of its size is unable again to descend. If this occurs to but one testis, the animal is termed a *monorchid or rig*, and if the other testis has been removed by castration, he will be lustful, but an uncertain stock-getter.

The immediate coats of the testis are a serous layer, *tunica vaginalis testis* and *tunica albuginea*. The latter is a dense fibrous mass, forming the stroma of the organ, sending fibres and septa inwards from an outer layer, to support and separate the different lobules of the gland, and to invest its larger vessels. It predominates at the upper part of the organ, where it is termed *mediastinum testis*. The testis consists of a number of long tubules curled upon themselves, forming the *lobuli testis*. Each lobule is isolated from its fellows, receives a vascular coating (by some anatomists termed *tunica vasculosa*), and terminates superiorly in a *vas rectum*. The vasa recta unite and form a plexus in the substance of mediastinum testis, at the anterior part of which they converge to form the *vasa efferentia*, which emerge from the testis proper, and, becoming very convoluted, (*coni vasculosi*), form the *posterior globus of the epididymis*.

The testicle, thus constituted, is a rounded body, elongated from before backwards, slightly flattened from side to side. It is firm to the touch, but on laying it open we find that it consists of a quantity of soft greyish matter. From its antero-superior part the **epididymis** commences and extends from before backwards, consisting of *an anterior and a posterior globus, and a body.* The *anterior or smaller globus* consists of the contorted vasa efferentia; they, however, unite and reunite, becoming larger in calibre, until they form but one tube, which, by its extreme convolution, forms the *body* and the *posterior or major globus*, from which, as the **vas deferens,** it passes up the posterior part of the spermatic cord towards the pelvis, where it becomes dilated on the supero-lateral part of the bladder, and, after passing beneath the prostate, opens into the urethra near its commencement.

The epididymis is connected to the testis by a double fold of visceral vaginal serous membrane, which is continued upwards to the internal abdominal ring surrounding the spermatic cord. In the anterior part of the cord runs the extremely convoluted spermatic artery with the plexus of spermatic veins or *corpus pampiniforme;* in the posterior part the vas deferens with the artery of the cord. *The walls of the sperm-bearing vessels or tubuli seminiferi* increase in complexity from their commencement to the termination of the vas deferens in the urethra. At first they consist of but basement membrane and epithelial cells, in which latter may be seen round bodies, which in their course to the vas deferens gradually acquire tails, and at length become free as **spermatozoa,** small, actively moving bodies, propelled by the vibration of an elongated cilium, running from the posterior part of the cell, giving it a "tadpole-like" appearance. The muscular walls of the vas deferens are thick, giving it its distinctively firm feel in its course through the spermatic cord.

The testis derives its nerves from the hypogastric plexus.

Vesiculæ seminales are pyriform sacs, resting upon the supero-posterior part of the bladder, in such a manner as to conceal the terminations of the ureters. They have but very thin walls, and each terminates anteriorly in a blind pouch, posteriorly in the duct which unites with vas deferens to form the ejaculatory duct, which runs to a special opening on the veru montanum. It presents a small amount of muscular structure in some parts of its walls, and is lined with mucous membrane. There is no doubt but that it secretes a special fluid to be added to the semen. It also may, to a certain extent, serve as a receptacle for retention of the semen until it is required for use.

The generative organs of the male produce the sperm-cell, those of the female the *germ-cell or ovum*, and without contact of these two cells reproduction cannot occur. The ovum is produced in a special organ, one of which is situated on either side of the spine in the lumbar region. These are the **ovaries or female testes**. They are smaller than the testes of the male and more rounded, and their external surface is generally lobulated in consequence of projection of certain vesicles which the organ contains. It is invested by peritoneum, being situated at the anterior part of the broad ligament of the uterus, and presents a white fibrous stroma, *tunica albuginea*, embedded in which are certain vesicular bodies lined with tessellated epithelium, and containing a quantity of albuminous fluid, in which floats the ovum. These cysts receive the name **Graafian vesicles**, and their lining membrane, which is thickest at that part with which the ovum lies in contact, is the *membrana granulosa*. The **ovum** *is a cell consisting of a cell-wall*, **zona pellucida**, *and cell contents;* the latter consists of a *nucleus* or **germinal vesicle**, and a *nucleolus* or **germinal spot**, floating in an albuminous fluid.

The Graafian vesicles are smallest at the centre of the organ, and gradually increase in size on extending towards the outer surface. By pressure they cause absorption of the stroma, and thus at length bulge against the peritoneal coat of the organ. Finally they rupture, thus discharging their contents into the Fallopian tube. This occurs when the animal is in "heat," as it is commonly termed, and thus the ova commence their process of development into the fœtus. After rupture the Graafian vesicles become filled with a mixture of blood and lymph, which undergoes contraction and organization until a stellate cicatrix, **corpus luteum**, is produced, which is much more persistent when its ovum has been impregnated than when it has undergone no further stage of development. These bodies form our main guide in post-mortem decision, as to whether an animal has ever been pregnant or not.

Attached to the ovary is the **Fallopian tube**, a trumpet-shaped organ, the anterior extremity of which presents a very irregular margin, which is much jagged, and is termed the *fimbriated extremity or morsus diaboli*. It is one of the divisions of this extremity which connects it to the ovary. It is perforated by a canal, which posteriorly gains the cavity of the uterus at a very small extremity in the extreme anterior part of the horn of that organ. Anteriorly it dilates, opening into a peculiar serous sac of which the ovary forms one boundary and which communicates on one side freely with the main peritoneal cavity. The *ciliated* epithelium with which it is lined is here replaced by tessellated epithelium. This is considered remarkable as an instance of a mucous membrane being directly continuous with a serous membrane, but the histological distinction between the two is insignificant. During sexual excitement the fimbriated extremity is supposed to grasp the ovary, approximating its anterior opening to the rupturing Graafian vesicle. The tubes are convoluted, and present muscular fibres of the unstriated class in their walls.

The **uterus** is a large organ which occupies the anterior part of the pelvis and the posterior part of the abdomen. Its position, however, varies with its pregnant or unimpregnated condition. Anteriorly it is bifid, presenting two rounded sacs, the *cornua or horns*. These meet at an acute angle in the mare; thus the equine uterus presents no *fundus* or space between the cornua. They open into the *body*, which extends backwards to terminate in the *neck*, the portion running for a short distance into the vagina to terminate at the os uterinum externum. The opening leading from the cavity of the body of the organ to the passage of the neck is *os uterinum internum*. The uterus is completely enveloped by peritoneum, being enclosed in a double fold of that membrane, which on either side extends to the supero-lateral parts of the abdomen, forming the *broad ligaments of the uterus*, on the free margin of which are the ovary, fallopian tube, and uterine horn. Between the two layers of this ligament the **round ligament of the ovary** runs from that organ to the extremity of the horn of the uterus, from which point another white fibrous (or muscular band) runs towards the internal abdominal ring, where it becomes lost, the **round ligament of the uterus**. The **ovarian artery**, after being given off by the posterior aorta, runs through the broad ligament to the ovary, where it breaks up, forming, according to some anatomists, a *tunica vasculosa* surrounding the organ. It corresponds to the spermatic artery of the male, than which it is much smaller. Through this ligament, the vessel corresponding to the artery of the cord in the male, the **uterine artery** takes a convoluted course downwards and backwards. It is much larger than the artery of the cord, and increases very considerably in size during pregnancy. The *muscular coat* of the uterus presents two layers of fibres; longitudinal externally, circular internally (it also has oblique fibres). These are of the unstriated order, but during pregnancy become developed into fibres with faint traces of striæ. The *mucous membrane* is of a reddish colour, and collected into rugæ in the unimpregnated uterus. It becomes highly vascular, and its epithelium much increased in quantity during pregnancy; for the blood of the fœtus passes to the chorion, a membrane lying in intimate contact with the vascular part of the mucous membrane of the dam, and in which the process of aëration of the blood ensues. (The excess of epithelium after parturition is thrown off as the **membrana decidua**.) In some parts, especially up against the fallopian tubes, the epithelium is *ciliated*, serving to waft the spermatozooa upwards. Around the neck of the organ the mucous glands increase in complexity, and secrete a more viscid matter; they are the *glandulæ of Naboth*. The muscular coat in the cervix uteri is very thick, thereby causing puckering of the mucous membrane longitudinally in this part; laterally it comes in contact with the muscular coat of the vagina, and posteriorly, covered by the mucous membrane, projects in a remarkable manner into the vaginal cavity "like the tap of a beer barrel," thus forming the *cauliflower excrescence* (*fleur épanouie of the French anatomists*), which projects more inferiorly than superiorly.

The **vagina** is the cavity placed immediately posterior to the uterus extending to the vulva, from which it is separated by the hymen. It is largest *centrally*, becoming *anteriorly* constricted in surrounding os uteri, and running to join the cervix; *posteriorly* smaller in joining the vulva; *superiorly* it is in contact with the rectum; *inferiorly* with the bladder. Its *anterior part* is covered with peritoneum, which laterally blends with the broad ligaments of the uterus, and sends off longitudinal folds; superiorly the *vagino-rectal*, inferiorly the *vagino-vesical*, folds to the rectum and bladder respectively. It presents also a *muscular coat* with longitudinal and circular fibres; in contact with this lies a quantity of *erectile tissue*, analogous to that forming the corpus spongiosum of the penis. It is lined by mucous membrane, with stratified squamous epithelium, and which anteriorly becomes puckered up to join that of the neck of the uterus; posteriorly forms a remarkable double fold, **hymen**, almost, but seldom completely, separating the vaginal cavity from that of the vulva. Generally lunate in the virgin, it is lacerated in the first act of coition by the penis of the stallion, and subsequently appears as an irregular fringe extending round the vagino-vulval opening, **carunculæ myrtiformes**.

The **vulva** leads from the vagina to the external generative opening, and presents coats similar to those of the vagina, with a much greater quantity of erectile tissue (**bulb of vulva**). Posteriorly the muscular coat becomes redder in colour, with its fibres striated and arranged in a circular manner around the external opening, forming the **sphincter vaginæ** (more correctly **sphincter vulvæ**). The floor of the vulva, just behind the hymen, about four inches from the lower part of the external opening, presents the **meatus urinarius externus**, much larger than that of the male, being the opening from a larger canal; between this and the hymen anteriorly is a valvular flap of mucous membrane.

The **external generative opening** is an elliptical fissure extending from above downwards. It terminates above in an acute angle, *superior commissure*, situated about two inches below the anus, the intermediate space being perineum. Below, it is more rounded, forming the *inferior commissure*. The opening is bounded laterally by the **labia**, elongated prominences consisting of the sphincter vulvæ, fat, and vascular structure covered externally by thin, hairless skin, which externally is continuous with the mucous membrane of the vulva, the latter becoming folded longitudinally to form the **labia minora or nymphæ**. At the inferior commissure is a triangular space, which may be exposed by separation of the labia, **fossa navicularis**, on which projects the **clitoris**, a small organ analogous to the penis of the male, and the seat of the pleasure experienced during sexual intercourse. It is surrounded by a mucous fold, **præputium clitoridis**, which covers its free extremity, and centrally presents a depression leading into a blind sac, which generally contains an odorous oily matter. This organ is fixed by two **crura** to the ischial arch, has two **erectores clitoridis** running to its substance from the same part, and presents a *corpus cavernosum*, a *corpus spongiosum, and a glans*.

The young animal at birth being incapable of supplying itself with nutriment, and even incapable of digesting ordinary alimentary matters, is for a time supplied with milk produced by the mammary gland of the dam. This function the mare shares in common with all the females of the higher order of vertebrated animals, **mammalia**. Situated between the thighs are two prominences, from the under part of which small finger-like portions project, having at their extremity three or four small openings. These are the **teats**, and they are covered by a continuation of the delicate skin investing the **mammary glands**. They are longest in the animal which has "given suck." The faschia superficialis abdominis, at the postero-inferior part of the abdomen of the mare, is reflected downwards to invest the mammary gland. It sends in septa between the different lobes, for this is a racemose gland, presenting a structure similar to that of the salivary glands. Its primary secreting saccules, lined by *spheroidal* epithelium, open into the primary ducts, which unite and reunite with those from other parts of the gland, until at length they terminate in a dilated cavity at the base of the teat, **galactophorous sinus**. This is divided into parts corresponding in number, and leading by ducts to the openings on the teats. The teats contain yellow elastic tissue, but no muscular fibre, for the milk is withdrawn by the suction action of the young animal. Around the openings the skin becomes continuous with the mucous membrane of the glands.

Milk consists of water with an albumenoid matter, *casein* (cheese), and *salts*, in solution. Floating in it are numerous fat-globules, which form the *cream*. The first milk produced after foaling contains a number of the epithelial cells of the gland in a state of fatty change, a number of fat-globules being connected together by the tough wall of the cell, forming the peculiar **colostrum-corpuscles**, for this milk is termed **colostrum (or beastlings)**. It has a laxative effect, and serves to remove from the alimentary canal of the newly born animal the **meconium** or *inspissated biliary matter* which has accumulated in the intestines while the animal was in the uterus, in consequence of the large supply of blood received by the liver during that period. Where this precaution is neglected, the meconium sometimes collects in large quantities of pellets, like sheep dung in the rectum, causing obstinate constipation.

The dissector may be fortunate enough to obtain a pregnant animal for dissection; he will then be enabled to study the **FŒTAL APPENDAGES.** By cutting through the walls of the uterus, he will expose the **chorion** lying in immediate contact with the uterine mucous membrane over its whole surface. It consists of an intricate network of blood-vessels arranged in a plexiform manner to come in apposition with the vascular layer of the uterine mucous membrane. While in these vessels, being separated from the blood of the dam only by layers of epithelial cells and of basement membrane, the blood of the fœtus is enabled to exchange carbonic anhydride for oxygen, sufficient in quantity to support his torpid life, for he moves but little, and his temperature is maintained by that of his dam in whose body he lies; thus two of the demands for oxygen in the blood which obtain after birth are absent in the fœtus, and his life consists more in active addition to existing structures, growth, than in disintegration of tissue with substitution of fresh matter, such as obtains in the adult. Blood from the internal iliac branch of the posterior aorta of the fœtus passes through the **umbilical arteries.** These vessels, which in the adult become impervious, degenerating into the *round ligaments of the bladder*, after becoming attached to their respective sides of the bladder gain the umbilical opening; here they unite and form one vessel, which runs outside the body of the fœtus to the chorion through the umbilical cord. The veins of the chorion unite and reunite to at last form two vessels which pass along the umbilical cord, and on gaining the umbilicus unite to form the **umbilical vein**, which runs in a forward direction along the floor of the belly between the peritoneal folds, forming the falciform ligament of the liver to which organ it runs without (in the colt) having any direct communication with the posterior vena cava through the **ductus venosus** as found in many other animals. It passes between two of the divisions of lobulus scissatus, where it breaks up, and its blood, after ramifying through the liver (which is very large in the fœtus), as in the adult, gains the posterior vena cava through the hepatic veins. It thus passes to the heart, and the major portion of it through the *foramen ovale* in the septum auricularum into the left auricle, where it meets with blood from the lungs, and with it through the left ventricle passes into the aorta, and from that into the anterior aorta, for the posterior aorta is mainly occupied by another current. The blood which gained the right auricle from the anterior vena cava, though a small amount of it passes through foramen ovale, for the most part having traversed the right ventricle, gains the pulmonary artery, from which some passes to the lungs, but a considerable portion through the **ductus arteriosus**, a vessel connecting the pulmonary artery to the posterior aorta. Through the posterior aorta some passes to the body of the fœtus and some through the uterine arteries to be reoxygenated in the chorion. **Foramen ovale** is an opening through septum auricularum just in front of the termination of the posterior vena cava; it is guarded by *a valve formed of a double fold of endocardium*. After birth it becomes obliterated on the left side, traces of it remaining on the right side, forming the *fossa ovalis* surrounded by a prominent yellow ring, *annulus ovalis*.

Immediately, but loosely, surrounding the fœtus in utero is the **amnion**. This is a thin membrane which presents some extremely convoluted vessels, ramifying over its surface, and serves to secrete a fluid in which the fœtus floats (**liquor amnii**). It is continuous with the skin of the fœtus at the umbilicus, becoming reflected over that portion of the umbilical cord nearest to this opening, and which is therefore termed the amniotic portion. Floating about in the liquor amnii generally are one or more brownish masses of soft substance. These are the **hippomanes**, and are supposed to be composed of nutritious matter. Sometimes they are attached to the inner surface of the amnion. In this country they are generally termed "*false tongues*," from their peculiar shape.

Investing the amnion, and lining the chorion, is a serous membrane, the **allantois**. Its *parietal or chorial portion* is continuous with its *visceral or amniotic portion* by a portion reflected over that part of the umbilical cord which extends from the amnion to the chorion. The amnion presents the characters of all other serous membranes, but it contains the *urine of the fœtus*, since the **allantoid sac** is brought into communication with the bladder of the fœtus (from which it was originally derived) by the **urachus**, a tube running through the amniotic part of the umbilical cord. The fœtus in utero, therefore, is surrounded by his own urine. The **umbilical cord**, therefore, is divided into two parts: The *allantoid portion* consists simply of two umbilical veins, and an umbilical artery surrounded by the allantois. The *amniotic portion*, in addition, contains the urachus, and those vessels which belong to the amnion. In both parts the vessels receive coatings of *embryonic cells,* and in the amniotic portion the vessels are arranged spirally, giving this part a rope-like appearance. Finally, we may notice that the *bladder of the fœtus* is larger in proportion than that of the adult, extending to the umbilicus, where it terminates in the urachus. That the *so-called vascular glands* are largest in the fœtus (in proportion); that the *kidneys* are lobulated; the *testes in the lumbar* region in the colt; and the *lungs*, dark red solid organs, for they have not yet been filled with air. When the labour-pains of the dam occur the chorion is ruptured, and the amnion with its serous layer is the first to protrude through the external opening, being vulgarly termed the "*water bag.*" On rupture of this the fœtus is expelled, and sometimes in falling ruptures the umbilical cord. The membranes follow at variable periods. On first feeling the chill external air the young animal gasps, and so draws air through the trachea, it permeates the air-vessels, and for the rest of his life the animal is an air-breather. Those communications between the several portions of the vascular arrangement of the fœtus, which are not required after birth speedily become impervious; if they are persistent they cause disease.

We may examine the attachments of some muscles underneath the loins.

Psoas magnus arises as far forward as the inner surface of the last two ribs; it is attached to the transverse processes of the lumbar vertebræ, and its fibres inferiorly join with iliacus in becoming inserted into trochanter minor internus. Arteria profunda femoris runs across the inferior tendon.

Psoas parvus is longer and thinner than psoas magnus. Its fibres from the lateral part of the bodies of the last three dorsal and the anterior lumbar vertebræ converge to form a tendon, which becomes inserted into the brim of the pelvis midway between the antero-inferior spinous process of the ilium and the acetabulum.

Quadratus lumborum is attached to the last rib, and to the transverse process of all the lumbar vertebræ running to the venter ilii near the crista.

Intertransversales lumborum are situated between the transverse processes of the lumbar vertebræ.

The **psoas faschia** covers the under surface of the psoæ muscles. It affords attachment to transversalis abdominis and obliquus abdominis internus, and from it in some subjects arise sartorius and cremaster muscle.

PART IX.—SPECIAL ANATOMY.

The Hind Extremity.

FROM an incision which the operator first makes along the central line of the quarters he dissects the skin back over the external surface of the quarter, and posteriorly as far as the incision originally made along the central and inferior line of the body towards the root of the tail. By so doing he exposes a wide-spread and rather thick layer of substance of an elastic fibro-adipose material, which covers the muscles of the quarter, and anteriorly blends with the lumbar faschia, while superiorly, in blending with its corresponding layer from the opposite side, it becomes firmly attached to the superior vertebral ligament in the sacral region. This **faschia of the quarter** inferiorly is imperceptibly blended with the faschia lata or the faschia covering the thigh, and on removing it we find that gluteus maximus and gluteus externus are attached to it to a considerable extent, whence care is required in the process to avoid injuring the appearance of the subject. By its removal we expose the

Gluteus externus, a V-shaped muscle, that is to say, one with two points of origin, which converge to a common centre of insertion. The anterior part we shall find arises in common with tensor vaginæ femoris from the inferior part of the antero-inferior spinous process of the ilium, while the posterior is attached to the spines and transverse processes of the second and third sacral bones. The muscular fibres of both these parts are more or less attached to the under surface of the gluteal faschia, and blend opposite the anterior tubercle of trochanter major, where they become, for the most part, tendinous, and from this they run to the trochanter minor externus, where they are firmly inserted. Over the anterior tubercle this tendon plays freely through the medium of a synovial bursa. Almost as far as the inferior part of the anterior extremity of this muscle

Tensor vaginæ femoris blends with its anterior part to a considerable extent. It arises from the inferior part of the antero-inferior spinous process of the ilium, and inferiorly forms an aponeurotic expansion, which blends with the faschia lata, and which slightly extends over the internal surface of the muscles in front of the femur. It becomes firmly attached to the external surface of the patella. By dividing and dissecting back these muscles we expose

Gluteus maximus, a large bulky mass of muscular substance, clothing the dorsum of the ilium. It arises from the tendinous structure of the longissimus dorsi by a projecting process, which grows narrower and thinner until it terminates opposite the last rib; from the antero-inferior and supero-posterior spinous processes and dorsum surface of the ilium as far down as the neck; also from the anterior part of the sacro-sciatic ligament. Inferiorly it is inserted by two heads; one runs to the anterior tubercle of trochanter major of the femur, it constitutes a broad tendon, which first passes over the superior smooth surface, from which it is separated by a bursa, and then becomes attached to a roughened ridge; the other contains muscular fibres, and after becoming inserted into the prominent posterior tubercle of this eminence, sends muscular fibres to the ridge running from it to trochanter minor externus. By careful division of this muscle inferiorly we may expose **gluteus internus,** which arises from the external surface of the neck of the ilium, and runs to the roughened internal surface of trochanter major; it crosses, and is attached to the capsular ligament of the hip-joint, and on the dry bones its superior point of attachment is marked by a transverse irregular ridge.

Triceps abductor femoris arises from the superior spinous ligament of the middle sacral bones, and from the sheath of faschia which invests the coccygeal muscles of the lateral part of the sacrum, blending with biceps rotator tibialis, from the sacral transverse process, and from the sacro-sciatic ligament; it receives a bulky mass from the tuberosity of the ischium, and after sending a somewhat irregular fibrous band to the posterior surface of the femur, midway between the two small trochanters (with ischio-femoralis), divides into three portions. The *superior* runs to the outer surface of the patella, the *middle* to the outer surface of the tibia and the outer straight ligament of the patella, blending with the faschia lata, the *inferior* is said to blend with gastrocnemius externus, but careful examination shows that it joins the white fibrous tissue layer, which arises from a ridge in common with that muscle, and with this runs to form that firm tendinous band into which plantaris passes, and which extends to the point of the hock.

Biceps rotator tibialis arises from the superspinous ligament and the transverse process at the posterior part of the sacrum, like the other muscles of the region, somewhat indefinitely. It forms a considerable portion of the fleshy mass of the quarter, and in its course down the limb receives a branch of equal size from the tuberosity of the ischium. Though thus reinforced, it forms a somewhat small tendon, which winds round the smooth internal surface of the tibia, where it is lubricated by synovia to reach the sharp ridge on the antero-superior part of that bone. From this tendon also a white fibrous band runs to the hock. This muscle is also termed *semi-tendinosus*. Posteriorly it is in contact with *semi-membranosus* or

Ischio-tibialis, the muscle which forms the prominence at the posterior part of the quarter. This muscle arises mainly from the tuberosity of the ischium, but superiorly sends a thin band in an upward direction to be attached to the sacro-ischiatic ligament and to the posterior sacral and the anterior coccygeal bones. Inferiorly it passes to the internal surface of the inferior extremity of the femur and of the head of the tibia, and the lateral ligament which connects them. It also sends a band of white fibrous tissue to unite with that of biceps rotator tibialis and run to the point of the hock. By removal of the three last-mentioned muscles we expose superiorly, up at the lateral part of the spine, the coccygeal muscles enveloped in a white fibrous tissue sheath. Below this the sacro-sciatic and the sacro-ischiatic ligaments, on the external surface of which are to be seen the gluteal and sacro-sciatic nerves, and the gluteal arterial branches which emerge from the pelvis through foramina in the ligament. At the infero-posterior part we see certain muscles winding round the neck of the ischium to become inserted into the trochanteric fossa. Having examined these muscles on the outer surface of the limb, we will now examine those which form the rounded fleshy mass on the inside of the thighs. On removal of the skin from the central line we expose gracilis, occupying the posterior part of this region, sartorius centrally, vastus internus anteriorly placed. Between gracilis and sartorius pass a large branch of the crural nerve and the **vena saphena**, to gain the femoral space. They run from the hock over the inner surface of the tibia and the thigh immediately beneath the skin.

Gracilis, supero-posteriorly, intercrosses its tendinous fibres with those of its fellow, and supero-anteriorly blends with the tendon common to the abdominal muscles; it is thus indirectly attached superiorly along the whole length of the pelvic symphysis. We have to cut through the tendinous structures at the point of junction of the two limbs for more than an inch before we reach the bone. Inferiorly this muscle forms an expanded tendon, which sends off a band passing to the white fibrous tissue band running to the hock, but mainly becomes inserted into the inner straight ligament of the patella, into the inner part of the head of the tibia, and into the faschia clothing the inside of the thigh, blending anteriorly with

Sartorius, a thin long ribbon-like muscle, which arises from the under surface of the psoas faschia, and from the antero-inferior spinous process of the ilium through the medium of iliacus, and is inserted by blending inferiorly with gracilis. It covers that portion of the crural nerve which intervenes between the brim of the pelvis and the femoral space.

The **femoral space** is an elongated space situated in this region, through which pass the femoral artery and vein, and crural nerve; and which also contains the **inguinal lymphatic glands** and some fat. It is bounded: anteriorly, by sartorius; posteriorly, by pectineus, and biceps adductor femoris inferiorly; externally by vastus internus; internally, by gracilis.

On removing the layer of muscles formed by gracilis and sartorius we expose supero-anteriorly vastus internus and rectus femoris; posteriorly ischio-tibialis and biceps rotator tibialis; centrally pectineus and biceps adductor femoris, between which run an artery, vein, and nerve.

A thick band of white fibrous tissue fills up the triangular space, which is bounded by the brim of the pelvis on either side from the pectinean tubercle to the symphysis. Around this the **pectineus** winds in a direction backwards to its insertion into a roughened line on the inner surface of the femur, commencing at the inferior extremity of the trochanter minor internus. To the under surface of the above-mentioned ligament is attached the common tendon of the abdominal muscles, and from each of its lateral parts a round tendinous cord runs to pass above the transverse ligament of the hip-joint, and becomes inserted into the head of the femur; it is the *pubio-femoral ligament*. Posteriorly placed to this is *a large vein* which runs from one external iliac vein in a transverse direction to the other, and which receives on either side in the male a branch from the penis, in the female from the mammary gland; by this vein (the transverse branch) pectineus is separated from the small head of

Biceps adductor femoris, which arises from the under surface of symphysis pubis. The large head of this muscle arises from the under surface of symphysis ischii; opposite trochanter minor internus, the two combine to form a single fleshy mass, the muscle then redivides; the smaller head becomes inserted into the posterior part of the middle and inferior thirds of the femur just below the inner small trochanter; the large extends obliquely over the posterior surface of the femur as far down as the *outer* condyle; the femoral artery and vein pass between its fibres, thus apparently dividing it into two heads, the lower of which blends with the insertion of ischio-tibialis, thus becoming attached to "the internal surface of the inferior extremity of the femur, and of the head of the tibia and the lateral ligament which connects them."

On removal of these muscles we expose the insertion common to iliacus and psoas magnus into trochanter minor internus.

Iliacus arises from the venter surface of the ilium, extending as far downwards as the pectinean tubercle, as far back as the transverse process of the first sacral bone to which it is attached. It is in contact with the inner surface of the capsular ligament of the hip-joint, to which it is attached, and blends with psoas magnus to become inserted as above mentioned into the inner small trochanter. Externally it is in contact with the origin of

Rectus femoris. This muscle combines with the two vasti to form that large fleshy mass which clothes the front of the femur, by some anatomists described as a single muscle under the term **triceps extensor cruralis**, since, being inserted below into the upper surface of the patella, they serve through the medium of its three straight ligaments to extend the tibia on the femur. Rectus femoris arises by two flat tendinous bands from the external and internal surfaces of the neck of the ilium; between these is a bursa with a mass of fat. It passes over ilio-femoralis at its attachment to the capsular ligament of the hip-joint, and opposite, about the centre of the femur, blends with the vasti.

Vastus externus arises from the antero-external part of the femur, extending as high up as trochanter major, opposite the middle third of the femur it blends with

Vastus internus, which is attached to the antero-internal part of the superior two thirds of the femur. Between it and vastus externus superiorly is the inferior attachment of

Ilio-femoralis, a small muscular band attached superiorly to the front of the neck of the ilium, below the heads of rectus femoris, passing over and becoming attached to the capsular ligament of the hip-joint, which it prevents from injury during movements, and becoming inserted for about two inches and a half along the upper part of the central line of the front of the femur. It thus corresponds exactly to scapulo-humeralis posticus of the fore limb.

In the attachments of the "triceps cruralis" to the patella are distinguishable several synovial bursæ, and between rectus femoris and vastus internus run the main branch of the crural nerve, and the inguinal artery and vein. These parts just described are supplied with blood by the **external iliac artery,** which arises from the posterior aorta, where it passes on to the underneath surface of the sacrum, about two inches before it breaks up to form the internal iliac arteries. Just at its origin it gives off the **circumflex artery of the ilium,** which runs forwards and downwards to the antero-inferior spinous process of the ilium, supplying in its course iliacus and proceeding on into the flank, where its terminal ramifications anastomose with those of the lumbar, intercostal, internal thoracic, and epigastric arteries. The **artery of the cord (of the male)** generally arises from the external iliac near its commencement, and runs to the internal abdominal ring, where it commences to supply the structures of the posterior part of the spermatic cord; in the female this artery is termed the **uterine**, it is very large and runs directly backwards to supply the uterus, being much enlarged in pregnant animals. The main trunk of the external iliac continues along the brim of the pelvis and about one and a half inches below the pectinean tubercle it breaks up, forming the arteria profunda femoris and the femoral artery. **Arteria profunda femoris** passes in a direction backwards, but gives off the epigastric artery, which runs forwards. It then passes between pectineus and the terminal portion of psoas magnus and iliacus above trochanter minor internus, then externally to ischio-femoralis, and so gains triceps abductor femoris, which it mainly supplies. It sends a long branch downwards with the sacro-sciatic nerve to supply biceps rotator tibialis and ischio-tibialis. The **epigastric artery** arises from arteria profunda femoris about half an inch from its origin, courses its way forwards over the inner surface of the femoral artery and thus gains the muscles of the flank, where its terminal ramifications blend with those of the numerous arteries supplying this region, while one of its branches, the **external pudic**, passes downwards, and after passing through the external abdominal ring, in the female supplies the mammary gland with blood, in the male the penis, breaking up to form the *anterior dorsal artery of the penis*, which runs forwards towards the glans, and the *posterior dorsal artery of the penis*, which runs backwards to anastomose with the branches of the internal pudic branch of the obturator artery.

The **femoral artery** from the neck of the ilium passes downwards over the internal surface of the insertion of iliacus, and here gives off the *inguinal branch*, which dips with the crural nerve between rectus femoris and vastus internus. It then passes over the inner surface of the femur, through the femoral space, giving off muscular branches forwards and backwards and the *medullary artery of the femur*, the largest medullary artery in the body. It courses its way between the insertions of pectineus and the short head of biceps adductor femoris on the one side, and the internal surface of vastus internus on the other, then dips between the two portions of the large head of biceps adductor. Thus it gains the posterior part of the femur, and after giving off a *large branch running directly backwards*, passes between the internal head of gastrocnemius externus and gastrocnemius internus.

After giving off the external iliac arteries, the posterior aorta becomes very small in calibre, and after running about two inches on the under surface of the sacrum, breaks up to form two **internal iliac arteries**, between which is sometimes seen the **middle sacral artery**. The internal iliacs diverge, running in a downward, outward, and backward direction, giving off the *artery of the bulb*, and terminate by breaking up into the *obturator, gluteal, and lateral sacral trunks*. The **artery of the bulb**, almost at its origin gives off the **umbilical artery**, which, pervious in the fœtus, in the adult degenerates into an impervious fibrous cord, which runs to the central part of the fundus of the bladder between the two folds of the broad ligament of that organ. It is termed the *round ligament of the bladder*, and in the fœtus serves to carry blood to the chorion for revivification. After giving off this vessel the artery of the bulb passes along the side of the pelvis, gives off *a large branch to supply the bladder and prostate gland*, and terminates in the bulb of the penis; in the female in the erectile walls of the vagina. It gives off *perineal branches*, which supply the posterior part of the rectum and the anus, anastomosing with branches from the lateral sacral and internal pudic arteries.

The **obturator** runs from its origin directly over the venter surface of the ilium to the anterior and external part of the obturator foramen, after passing through which it breaks up into three very indefinite branches: *pubic* runs forwards, *ischiatic* downwards, and the *internal pudic* to the bulb of the penis, breaking up in the erector penis muscle, and anastomosing anteriorly with the posterior dorsal artery of the penis, given off by the external pudic branch of the epigastric, posteriorly with the artery of the bulb. While passing over the surface of the ilium, the obturator gives off a branch which passes over the brim of the pelvis just above the origin of rectus femoris, and passes mainly between that muscle and vastus externus; it is the *arteria innominata*.

The **gluteal artery,** immediately after its origin, passes through the foramen situated between the posterior margin of the ilium and the anterior margin of the sacro-sciatic ligament. It then breaks up, sending large branches to the gluteal muscles, and among others a remarkable one running along the lateral margin of the sacrum between two folds of the ligament. Thus the muscles on the quarter are supplied with blood.

The **lateral sacral artery** passes from the terminal portion of the internal iliac in a backward direction along the outer margin of the under surface of the sacrum; thus the artery tends to approach its fellow. It terminates posteriorly in giving origin to *vessels running to the tail (lateral coccygeal arteries).* Superiorly it gives off *small branches to the spinal cord,* which pass through the subsacral foramina and *medullary arteries to the sacral bones.* Inferiorly it gives off *hæmorrhoidal branches to the rectum,* which anteriorly anastomose with the terminal branches of the posterior mesenteric, posteriorly with the artery of the bulb and its branches.

The **middle sacral** when present runs along the central line of the under surface of the sacrum, and terminates in giving rise to the *middle or inferior coccygeal artery.*

The **femoral vein** exactly corresponds to the femoral artery; just above the trochanter minor internus it receives the vena saphena (which we have already seen coursing along the inner surface of the limb from the hock, subcutaneously placed). Also in front of symphysis pubis it receives the transverse branch already noticed. It terminates superiorly in forming the **external iliac vein,** which unites with the **internal iliac vein** (which brings blood from the parts supplied by the artery of the same name) to form the **common iliac vein,** a vessel of about three inches in length, which combines with its fellow under the last lumbar vertebra, above the posterior aorta, to form the commencement of posterior vena cava. This vessel receives the **circumflex vein of the ilium**; it is to be found between the external and internal iliac arteries. The posterior extremity is supplied by four large nerves in addition to small sacral branches and the sympathetic system; these are the crural, gluteal, sciatic, and obturator.

Like all the other **spinal nerves,** those in the lumbar region, on emerging from the spinal canal through the intervertebral gaps, break up into superior and inferior branches. The *superior branches* pass upwards to supply longissimus dorsi and the other muscles in the lumbar region of the back; the *inferior branches* break up, sending some fibres to the abdominal muscles running over transversalis abdominis; others to the psoæ muscles; others from the second, third, fourth, and fifth to form the **crural nerve**, which passes along the brim of the pelvis, and opposite symphysis pubis breaks up, its main portion running between rectus femoris and vastus internus to supply the muscles clothing the front of the femur, a small, long branch (saphenous nerve) passing downwards with vena saphena subcutaneously as far as the hock. From the inferior branches of the third and fourth lumbar nerves the **obturator nerve** arises. It takes the same course as the artery of the same name, and supplies the structures on the external and internal surface of the obturator foramen and ligament.

The **sacral nerves** differ from the other spinal nerves in that they run in an oblique direction outwards and backwards from their origin, thus forming that arrangement in the terminal portion of the spinal or vertebral canal to which the name *cauda equina* has been given. They also divide into superior and inferior branches prior to leaving the canal. The *superior branches* pass in an upward direction through the supersacral foramina to supply the anterior portion of the coccygeal muscles and the upper parts of the muscles of the quarter; the *inferior branches* run to supply the muscles of the quarter, fibres from the first sacral, and the last lumbar nerves, combining to form the **gluteal nerve**, which passes through the foramen at the anterior part of the sacro-sciatic ligament between it and the ilium, thus gaining the external surface of the ligament, where it supplies the muscles of the quarter. The inferior branches of the last lumbar, with the first three sacral nerves, combine to form the

Sciatic nerve, the largest in the body, which, by passing with the gluteal, gains the external surface of the sacro-sciatic ligament, over which it runs on the inner surface of triceps abductor femoris over the neck of the ischium, between the cotyloid cavity and the tuberosity of the ischium, and down as far as the superior extremity of the gastrocnemii muscles. In its course behind the femur it is accompanied by two roots of vena profunda femoris, and a branch of arteria profunda femoris. Inferiorly it breaks up to form the *anterior and posterior tibial, and popliteal nerves*. In its course down the limb it gives off fibres to all the large muscles among which it runs. By removal of the above-mentioned structures we are enabled to see

Ischio-femoralis, which arises from the under surface of the neck of the ischium, and becomes inserted with the tendinous band of triceps abductor femoris to the posterior surface of the femur midway between the two small trochanters; it is triangular and fleshy; also the four muscles which become inserted into the fossa behind trochanter major are thus exposed; these are

Obturator externus, which arises from the under surface of the obturator ligament and the margins of the obturator foramen. It resembles a number of small muscles united together.

Gemini vel ischio-trochanterius, composed of two very similar united portions, which arise from the under surface of the neck of the ischium in front of ischio-femoralis.

Obturator internus, which arises within the pelvis from the upper surface of the obturator ligament, is superiorly covered by a reflection inwards of the sacro-sciatic ligament and is also attached to the margins of the obturator foramen. This muscle combines with

Pyriformis, arising from the posterior part of the venter ilii, as high up as the transverse process of the first sacral vertebra. The tendon common to the two gains exit from the pelvis by passing between the sacro-sciatic ligament and the neck of the ischium in winding round the bone, where it is lubricated by synovia; it then blends with obturator externus and gemini in passing to their common insertion. The limb may now be removed from the trunk by sawing through the middle of the femur transversely. The gastrocnemii muscles may then be examined.

Gastrocnemius externus arises by two heads from the ridges bounding the fossa on the postero-inferior part of the femur inclined to the outer side, one from the inferior third of the external surface, the other from the inferior third of the posterior surface of that bone. The external head is covered by a *broad aponeurotic band*, which runs to form a sheath for gastrocnemius externus, after receiving a band from the aponeurosis of gracilis. It then meets a funiculus running from gastrocnemius internus, and forms a loop through which plantaris runs. This band, plantaris, and gastrocnemius externus, together become inserted into the point of the hock, after playing over a bursa. About opposite the superior part of the inferior third of the tibia the tendon of gastrocnemius internus, which has been hitherto placed in front of the externus, winds *from within outwards* around the externus tendon, on which it plays through the medium of a synovial sheath.

Plantaris is a fleshy muscle, apparently too long for its situation, running from the superior part of the fibula to become inserted as above described into the superior part of os calcis. The aponeurosis of the lowest head of triceps abductor femoris sends downwards a broad band of white fibrous tissue; some of the muscular fibres of biceps rotator tibialis terminate in tendinous structure, which blends with the above; a little lower down this band receives fibres from gastrocnemius internus. About an inch and a half above the point of the hock it bifurcates, one division running on to the inner, the other to the outer surface of the point of os calcis, from which they pass to the dilated portion of the tendon of gastrocnemius internus, which forms the cap of the hock, which they serve to prevent from dislocation, being assisted by some ligamentous fibres which run directly from the cap to the bone.

Gastrocnemius internus arises from the fossa at the postero-external part of the femur, between the two heads of the gastrocnemius externus. It passes over the posterior part of the stifle joint, and after sending some tendinous structure to the fibrous band running to the hock, becomes tendinous, winds round the tendon of the external gastrocnemius muscle from within outwards, and then, widening, forms a cap for the point of os calcis, which is retained in position in the above-mentioned manner by two fibrous bands, and which plays over a large synovial bursa; continuing downwards it is termed the perforatus tendon, and exactly corresponds to that of the fore extremity.

Popliteus is attached inferiorly to a triangular roughened surface, situated on the inner part of the postero-superior region of the tibia; its fleshy fibres converge, forming a tendon which runs in an outward direction over the surface of the capsular ligament of the stifle-joint within its external lateral ligament (lubricated with synovia), to become inserted into a groove between the outer condyle and outer ridge of the femur.

Flexor pedis perforans also arises from the posterior surface of the tibia, but is externally placed, being also attached to the posterior margin of the fibula; it becomes tendinous above the hock, passes over the concave internal surface of os calcis, where it is lubricated with synovia; opposite the superior third of the large metatarsal bone receives the tendon of flexor pedis accessorius and the subtarsal ligament, which is not so strong as the subcarpal. Below this it comports itself in a manner corresponding to the similar tendon of the fore limb.

Flexor pedis accessorius arises from the posterior surface of the tibia, between the attachments of popliteus and the perforans; it becomes tendinous opposite the inferior third, and passes through a groove situated just behind the inner malleolus of the tibia, where it is bound down by a continuation of the posterior annular ligament, and lubricated with synovia, having a sheath which extends obliquely downwards and backwards over the lateral part of the hock. It terminates in becoming blended with perforans, opposite the superior third of the large metatarsal bone. On the external surface of the limb we see three muscles.

Peroneus arises from the outer surface of the fibula, being continuous superiorly with the outer lateral ligament of the stifle-joint. Its fibres converge in a bipenniform manner to form a small tendon, which passes through the groove found on the external malleolus of the tibia. It is here lubricated by synovia, and, enclosed in a synovial sheath, passes within the annular ligament in an oblique direction forwards, joining extensor pedis tendon with extensor pedis accessorius at the inferior part of the superior third of the large metatarsal bone.

Flexor metatarsi arises by a round tendon common to it and to extensor pedis from the depression between the outer condyle and the outer ridge of the femur. This tendon, after becoming attached to the capsular ligament, passes in the groove between the spine of the tibia and the head of the fibula, where it plays over a synovial bursa; it then gives off extensor pedis, and receives numerous muscular fibres from their attachment in the external channel of the tibia. The tendon, however, continues as a white fibrous band to the front of the hock, where it breaks up to become inserted into several of the small bones of that joint, while the tendons of the muscular portion pass between its divisions, thus completing its inferior attachment to all the small bones of the hock, directly or indirectly, and to the upper part of the three metatarsal bones. With one of these tendons the middle band of the anterior annular ligament of the hock blends.

Extensor pedis, arising from the originating portion of the tendinous part of flexor metatarsi, as above indicated, passes down the front of the hock, and through its three anterior annular ligaments, lubricated by synovia. After receiving the tendons of peroneus and extensor pedis accessorius opposite the inferior part of the superior third of os metatarsi magnum, it comports itself as does the extensor pedis of the fore limb.

Extensor pedis accessorius arises from the front of astragalus below its articular ridges, and joins extensor pedis, which covers it from its origin, being slightly visible between it and peroneus.

The terminal portion of the femoral artery passes between the two heads of gastrocnemius externus, and then between its internal head and the head of gastrocnemius internus. It sends off *a branch which runs downwards between the two gastrocnemii,* and then passes between the condyles of the femur to break up behind the stifle-joint, forming the *anterior and the posterior tibial arteries.*

The **posterior tibial artery** runs down the posterior surface of the tibia, at first between popliteus and the bone, subsequently between flexores pedis perforans and accessorius. It becomes superficial at the supero-internal part of the hock, and here makes its *sigmoid flexure,* by which, in terminating, it anastomoses superiorly with the arterial branch between the gastrocnemii, inferiorly with the recurrent branch of the metatarsal artery. It supplies the muscles and skin at the posterior part of the tibia, and the *medullary artery* of that bone at the inferior part of its superior third.

The **anterior tibial artery** runs in a forward direction from its origin, and passes between the tibia and the fibula, and thus underneath peroneus to gain the front of the limb, down which it passes underneath flexor metatarsi and extensor pedis, which it supplies. At the front of the hock it passes through an oblique groove at the outer part of the superior extremity of the large metatarsal bone, and thus gains the channel between this bone and the external small metatarsal, in which it runs subcutaneously, so that the pulse may be taken here. It passes between the inferior extremity of the small bone and the large bone to gain the posterior part of the inferior third of the latter, where, after giving off the recurrent branch, it bifurcates in passing between the divisions of the superior sesamoideal ligament, forming plantar arteries analogous in situation and distribution to those of the fore limb. The **recurrent branch** passes up the back of the os metatarsi magnum, protected by the inner small metatarsal bone, and at the inferior part of the superior third gives off *the medullary artery of the large metatarsal bone*. It terminates at the supero-internal part of the hock, by anastomosing with the posterior tibial at its sigmoid flexure.

The **superficial metatarsal vein** runs on the inner surface of the limb in the groove between the large and inner small metatarsal bones. At the front of the hock it bifurcates; one branch runs on to the internal surface of the tibia, forming the commencement portion of the vena saphena, the other, by uniting with a large branch running from between the bones of the hock, forms the **anterior tibial veins**, generally two in number, one of them being enlarged, the *varicose vein of the thigh*, they pass with the artery of the same name. The **deep-seated metatarsal veins** divide superiorly; one division forming the *interosseous branch*, which passes between the bones of the hock to assist in forming the anterior tibial, the other passing through the groove of the calcis to form the **posterior tibial veins**, which accompany the arteries of the same name up the back of the tibia.

Just above the commencement of the gastrocnemii muscles the sacro-sciatic nerve terminates in breaking up to form the *anterior and posterior tibial, and popliteal nerves*. The **popliteal** proceeds down the back of the limb, and dips between popliteus and the flexors, which it supplies with nerve force. The **anterior tibial** passes over the external head of gastrocnemius externus and peroneus to gain the muscles in front of the tibia. It thus presents a marked difference in its course as compared with that of the artery and vein of the same name. It is sometimes divided at its superficial part by the operation of neurotomy. To expose it, only skin, faschia, and aponeurosis of triceps abductor femoris require to be divided. The **posterior tibial nerve**, dipping between the internal head of gastrocnemius externus and gastrocnemius internus, follows the course of its corresponding artery over the back of the stifle-joint, and running down the posterior surface of the femur between the gastrocnemius and perforans, just above the hock, divides to form the **external and internal metatarsal nerves**, which, after passing with the perforans tendon over the inner surface of os calcis, separate and take a course, and give off branches exactly as the metacarpal nerves do in the fore limb.

In the space just above the heads of the gastrocnemii muscles lie the **popliteal lymphatic glands**. It will thus be seen that the hind limb below the hock very closely resembles the fore limb below the knee; besides the differences between the pedal and the cannon bones of the fore and hind limb, the hoofs differ, those of the hind limb being smaller, more upright, and having their ground surface much longer than broad, and more concave, with thicker soles than those of the fore.

The coccygeal muscles require a passing notice.

Levator coccygis arises from the lateral part of the sacrum between the superior spinous and transverse processes, blending anteriorly with longissimus dorsi. It passes backwards, and its several fasciculi become attached to the rudimentary transverse processes of the coccygeal vertebræ. It is covered by a layer of faschia.

Curvator coccygis arises from the postero-lateral and inferior part of the sacrum, and its fibres and fasciculi run along the lateral part of the tail; they are separated from levator coccygis anteriorly by faschia.

Compressor coccygis is a broad square muscle, which extends from the inner surface of the sacro-sciatic ligament to the transverse processes of the two or three anterior coccygeal bones.

Depressor coccygis is attached to the under surfaces of the anterior coccygeal bones, its fibres pass downwards towards the anus and become confounded with those of levator ani.

The tail is supplied with blood by the *lateral and middle coccygeal arteries*, with nerve force by *coccygeal branches* from the posterior part of the spinal cord.

PART X.—SPECIAL ANATOMY.

The Foot.

CONSIDERED zoologically the foot of the horse comprehends all those structures placed below the inferior extremity of the tibia; the point of os calcis corresponds to the heel of the human subject. From the tarsus the metatarsals run downwards; their normal number in mammalia is five, in the horse, however, we find but three, for the inner digit is absent and the two central ones coalesce. The outer of the three in the fossil forms of the horse present hoofs, and other structures found in the existing forms only in connection with the double central digit. In some cases even in the present day, however, we find a rudimentary hoof produced below one of the metacarpal or metatarsal splint bones, in rare cases; generally these bones terminate inferiorly in a slight bulbous enlargement. It is, then, with the double central digit we have here to deal; its composition of two similar parts is shown by its symmetry, and by a depression well marked in some cases at the antero-inferior part of the toe of os pedis. Again, in all cases the frog is divided into two similar portions by a deep fissure, the cleft, and often we see a projection on the inner surface of the hoof opposite the depression in the toe of os pedis just at the point of junction of the wall and the sole of the hoof. That portion commonly known as the foot, consists of a horny box with its contents. The horny box corresponds to the nail of the toe of the human subject with the thickened skin on the sole of the foot, while the contents comprise certain bones with their appendages covered by a layer of highly vascular tissue, which corresponds to the true skin, and which produces the horn, which is analogous with the epidermis or cuticle. The **HOOF** is separable from the other structures by maceration by a continuation of this process it may be separated into minor parts, we generally see it as a horny box presenting an opening above. It is composed of epithelial cells, being (as before observed) merely epidermis so modified as to form the substance commonly known as horn. **Horn** consists of tubular processes similar to hairs united by softer epithelial cells, the latter forming a matrix for the fibres and, by fixing them firmly together giving the horn its consistency. The horn of the hoof presents two parts with different characters. The *horny frog* is much softer in consistency and more elastic than the remainder of the hoof, and resembles in nature and structure (as pointed out by Professor Gamgee) the foot pad of the dog. It owes its peculiarities to the delicacy of its horn fibres, which are finer than those of most other parts of the hoof, and which are produced by the delicate papillæ of the sensitive frog. The hoof consists of several parts agglutinated together in the same manner though less firmly than the individual fibres. These are the wall (with the bars), sole, and frog. The frog and the sole are separable by maceration, but the bars are mere reflections forward and inwards of the wall, and therefore remain firmly fixed to that part. The **Wall** is all that portion of the hoof which is visible from before and from the sides when the foot rests upon its plantar surface. It is artificially divided into six parts, three on each side. The *anterior or toe* centrally lies in connection with its fellow, externally is continuous with the *quarter or middle part* at its anterior part, while the posterior margin of the quarter is continuous with the *heel*, from the posterior margin of which the *bars* run in a direction forwards and inwards for about two inches, being separated from the frog by the **commissures**, terminating anteriorly in a point. *In all parts the fibres of the hoof run obliquely downwards and forwards.* The toe and the posterior margin of the heel have the same direction, the latter being not half so high as the former in a well-formed hoof; their superior extremities are connected by a thin ridge running obliquely downwards and backwards, their inferior margin is horizontal. The *internal surface of the wall* superiorly presents a concave surface extending obliquely downwards and backwards about half an inch in depth, growing broader posteriorly in becoming inflected forwards to form the crust of the bars which blend with the horny sole anteriorly. It presents a number of deep small foramina, into which the papillæ of the vascular coronary secreting substance pass. The wall is divided into *crust and horny laminæ*; the **horny laminæ** are leaf-like plates projecting from the inner surface of the crust, taking a direction corresponding to the fibres of the wall, commencing superiorly at the inferior margin of the above mentioned perforated surface. They are longest at the toe and become insensibly merged in the wall at the anterior extremity of the bars. Between each pair of these in the fresh subject for the greater part of their length lies a sensitive lamina, and the surfaces of the horny laminæ present foramina into which pass papillæ from the sensitive counterpart. The *outer margin* of each lamina is continuous with the crust, the *inner* is free. The *superior extremity* terminates in a point in blending with the above described concave surface, the *inferior* terminates on the ground surface at *the line of junction of the wall with the sole*, at which part the outer laminated edge of the sole fits between the horny laminæ as the sensitive laminæ do in other parts; the extent of this junction depends upon the variable thickness of the sole. The horny laminæ are produced by the sensitive laminæ, and the crust by the coronary secreting substance; in consequence, therefore, of the constant additions to its thickness the wall has a tendency to greater bulk in its lower than in its upper part. That portion of the wall situated at the extreme superior edge, which is secreted by the line of vascular substance which presents very minute papillæ and seems to be the first stage of change in the character of the skin, thus presents fibres more delicate than those of the rest of the wall, corresponding to those of the frog, extending for but a short distance downwards, about half an inch. It is narrowest anteriorly and becomes deepest at the heels in blending with the frog. This is the **coronary frog band** of Bracy Clark (the original specimen of this by Bracy Clark in the museum of the Royal Veterinary College is much too thick anteriorly, where a small portion of coarser substance has been left). The **wall** is thickest at the toe, it gradually decreases in thickness towards the heels, and again increases slightly at the extreme posterior part, where it is inflected forwards to form the bars. In no part is the wall so thin as the coronary frog band at the extreme superior part. The angle the toe of the hoof should make with the ground has been stated as 50° in a good foot, but it is influenced by various causes, as breed, age, period since last shoeing, &c. Low bred horses have a small angle, in better bred horses, mules, and asses it is less acute. The obliquity decreases towards the quarter and is reversed at the heels, where the posterior line of the plantar surface is in front of the perpendicular let fall from the extreme posterior part of the coronet. The obliquity is downwards and outwards at the bars. The inner quarter is more upright than the outer, and the wall on the inner side, taken as a whole, is thinner and, therefore, more readily yields to superincumbent weight.

The **horny sole** presents two surfaces and two margins, which posteriorly form four projections. It is crescent shaped, having its *outer circumferent margin* attached to the inner surface of the inferior margin of the wall. Its *postero-internal part* is fitted into the horny laminæ of the bars, and continuous with the crust of the bars, while the *central part of the posterior margin* is occupied by the horny frog, which forms a wedge-shaped projection into it extending to within about two inches of the toe. The *inferior surface* is concave, varying in degree in different feet, but it presents a general slant from the outer margin upwards to the inner. It is covered by scales of horn, for it consists of a number of short coarse fibres matted together, which, after attaining a certain length, desquamate regularly. The union of the sole with the wall is by projection of its substance between the horny laminæ, and union of the two by connecting epithelial cells, forming a marked line of junction, softer in consistency than the rest of the hoof and more friable. It will be observed that the fibres of the sole run like those of the crust, obliquely downwards and forwards in all parts. The union with the anterior extremity of the frog is simply by epithelial cells. The *upper surface* of the sole is convex, an exact counterpart of the lower, presenting foramina, into which the papillæ of the sensitive sole project, and which are most marked centrally.

The **horny frog** is situated on the middle line of the hoof at the posterior part. It is a wedge-shaped body, having its base posteriorly, its apex anteriorly placed, extending to within two inches of the toe of the foot. It presents two *lateral margins*, which anteriorly are attached to the horny sole, posteriorly to the horny bars, and *at the posterior angles* are continuous with the coronary frog band. The posterior margin becomes very thin, blending with the epidermis of the skin of the heels. The frog bulges on the ground surface of the foot, presenting anteriorly a rounded angle or *apex*, and posteriorly, at the *base*, two rounded prominences termed the *bulbs*, separated by the *cleft*, which extends more than half the length of the frog in a forward direction, being widest posteriorly, terminating in a point anteriorly on the ground surface. The bulbs of the frog are separated from the bars by the *commissures*, deep fissures readily noticeable. The upper surface of the frog is the exact counterpart of the lower, presenting centrally a peculiar high prominence, terminating superiorly in a rounded point, which fits into the cleft of the fibrous frog (covered by its vascular coat). This is termed the **frog-stay or peak**, and on either side of it lies a concavity, widest posteriorly, by its elevated outer margin blending with the bars and sole. The two concavities run together anteriorly, terminating in a point. This surface presents small foramina for the delicate papillæ of the sensitive frog, least marked on the most prominent portions. The fibres of the frog run obliquely downwards and forwards. The **colour of the hoof**, in general, depends upon that of the sensitive structure by which it is produced; thus, in horses with white coronets it is generally white, but any coloration of the coronet is marked by corresponding colour of the hoof below this point, extending from above towards the ground surface, varying in length in direct proportion to the superficiality, or otherwise, of the pigmentary deposit.

The **hoof of the fore foot** is marked by the roundness of the plantar surface, broken posteriorly by the space for the frog; also by the obliquity of its walls, which are thickest at the toe, and decrease in thickness towards the heels. The distances from the centre of the toe to the extremity of each heel should be equal, and also equal to the broadest transverse diameter. The inner half of the sole is generally smaller and narrower than the outer; the inner quarter is the most upright, and is thinner in the wall than the outer.

The **hind hoof** is ovoid, having its long diameter from behind forwards. It is wide at the posterior part of the quarters, while their anterior part is rather flat. The hoof in general is more upright than the fore; the thickest horn is at the quarters, and the sole is very concave, the frog being smaller and narrower. The heels also are longer than those of the fore foot.

The **sensitive structures of the foot** are generally exposed by heating the foot in a forge fire, removing the heated horn from the outer margin of the plantar surface of the foot by means of a drawing knife, as far down as the sensitive structures, and wrenching the wall and sole thus separated off by means of pincers, the foot having been fixed in a vice. The frog will be removed with the sole. Some art is required in the operation, slow traction being preferable to sudden jerks. By this means we expose a red and highly vascular structure, investing the digit from the coronet downwards. Opposite the centre of the external surface of the lateral cartilage, extending round the foot anteriorly and posteriorly, is a line marking where the skin commences to change its characters. The epidermis at first becomes soft and light in colour, and then modified to form the hoof, some hairs from the coronet being embedded in the white cheesy upper rim of the hoof. The ordinary skin terminates in a regular and well-marked line, light in colour, covered with fine papillæ, and growing wider at the posterior part, becoming blended with the bulbs of the sensitive frog. Below this, extending in a downward direction for about half an inch, is the

Vascular layer of the coronary secreting substance. It gradually decreases in size in a direction backwards, more so on the inner than on the outer side, and at the posterior part, against the bulbs of the frog, it terminates in an acute point, being thickest just before this, in consequence of its inflection forwards to join with the vascular sole, towards which it runs, being here about two inches in length, a quarter of an inch in width, separating the sensitive laminæ of the bars from the sensitive frog. It differs slightly from the above description in the hind foot, in accordance with the different shape of the hoof. The vascular or sensitive coronary secreting substance is covered with papillæ, which are minute superiorly, but inferiorly become very marked. In some cases a few may be seen running from the upper part of the **vascular laminæ,**

which are thin plates covering the remainder of the antero-lateral surface of the basement structure of the foot, running from above downwards, at the anterior part meeting the ground surface at right angles, becoming oblique laterally. This obliquity from behind forwards and downwards increases towards the heels, the laminæ also becoming shorter in proportion. At the heels this layer is reflected forwards, slightly inwards, for about one and a half inches between the heel of the sole and the anterior inflection of the vascular coronary secreting substance, anteriorly and inferiorly blending with the sole, terminating in an ill-defined point, for here we see a useful indication of the *nature of the laminæ*, the fact that they are linear rows of papillæ united together for a specific purpose being here demonstrable. By careful examination we may see *secondary papillæ* extending horizontally into the interspaces between the laminæ from the surfaces of the layers. These laminæ are 500 to 600 in number; they are smallest superiorly, where they commence in blending with the coronary secreting substance; inferiorly they blend gradually with the sensitive sole.

The **sensitive sole** covers the whole of the plantar surface of os pedis, with the exception of a small wedge-shaped portion running forwards from the centre of the posterior crescentic margin, which is occupied by the apex of the sensitive frog. At the *anterior part* also, centrally, we find in most feet a marked notch, so that the sensitive sole is narrowest centrally, and from this gradually increases in width to the anterior part of the laminæ of the bars, where it bifurcates; the *outer division* terminating in a point at the inflection of the laminæ; the *inner division* blends with the termination of the coronary secreting substance. The whole surface presents papillæ, which around the outer margin are small and continuous with the inferior extremities of the sensitive laminæ. They gradually increase in size to the central line and then again decrease, being absent at the line of junction of the frog with the sole. The apex of the frog extends to within an inch of the toe of the sole.

The **sensitive frog** covers the external surface of the fibrous frog, and consequently presents two prominent ridges running from behind forwards, obliquely inwards, meeting at about the centre, and continued on to form a very acute angle at the apex. These are separated by the *cleft*, and on either side laterally blend with the sensitive sole. On the ridges are very marked papillæ, which gradually decrease in size in extending from the prominences until they are absent at the line of junction with the sole. There is a slight line caused by diminution in size of the papillæ, mapping out the boundary of the *commissures*. It may be noticed that the papillæ of the frog are smaller than those of the sole and of the coronary secreting substance. The **vascular horn-secreting layer** in all parts consists of loose areolar tissue, with very numerous vessels ramifying through it. The vascular sole is connected to the bone by loose areolar tissue, in which is situated a plexus of veins, and the **solar arteries** running from the artery of the frog outwards in a radiating manner to the **circumflex artery of the toe**, which runs round the outer circumferent margin of the plantar surface of os pedis. The sensitive laminæ are connected by denser areolar tissue, which attaches them to the antero-lateral surface of os pedis, where they cover the *lateral laminal arteries* and the branches they send upwards to the superficial coronary artery, and downwards to the circumflex artery of the toe. The sensitive layer of the coronary secreting substance is firmly attached to a somewhat dense band of substance **(fibrous coronary band)** which extends round the coronet. This is deepest at the toe, where it is attached to extensor pedis tendon at its insertion; laterally it rests on the centre of the lateral cartilage (external surface), and posteriorly it terminates in blending with the **fibrous or fatty frog.** This is an elastic substance, occupying the centre of the posterior part of the foot. It consists of a quantity of yellow elastic tissue arranged in fibres, which interlace in all directions, but the largest of which take a direction obliquely from above downwards and forwards. In the meshes of these fibres is a quantity of substance apparently of a fatty nature, whence this organ has received the name of fatty frog. It is a wedge-shaped body with two surfaces, a base and an apex. The *superior surface* lies in contact with the under surface of the expanded perforans tendon, with the intervention of a thin fibrous layer. It extends forwards on to the plantar surface of os pedis at the posterior crescentic margin at its centre, where it forms a triangular projection. Its *base* becomes blended posteriorly with the corium of the skin of the heels, and laterally receives the termination of the coronary fibrous band. The *apex* extends to within about one inch of the toe of the os pedis. The *lateral parts* superiorly are separated from inner surface of the lateral cartilage by means of a plexus of veins; inferiorly are firmly attached to the inflections of the cartilages, the transition between the two structures being gradual. The *inferior surface* presents a figure analogous to that of the under surface of the horny frog, which is moulded upon it with the intervention of the vascular frog. It therefore presents *two bulbs* separated from each other by a *cleft*, which form the *base*, and which converge anteriorly, and are continued forward to terminate in a point, the *apex*.

We now come to the basement structure of the foot, the several parts composing which we must now examine. The bones in the foot are, *os pedis, os naviculare,* and the *inferior extremity of os coronæ.*

OS CORONÆ, though broader than long, and possessing no medullary canal, is generally considered to be a long round bone. It really consists but of two extremities. Its *superior extremity* articulates with the inferior articulatory surface of os suffraginis, and thus presents a smooth surface coated with articular cartilage, consisting of two shallow concavities, the inner of which is the largest. These are separated by a slight prominence, extending in an antero-posterior direction ; the articulatory surface taken as a whole is concave from before backwards, and the prominence terminates posteriorly in a projecting point. The *inferior extremity* articulates with the os pedis and os naviculare; thus it presents a smooth surface which is broad posteriorly, and at the anterior part terminates in a point, to which the coronal process of the os pedis is adapted. The surface presents two prominences separated by a shallow groove; the inner division is slightly the largest. The *posterior surface* of the bone superiorly presents a prominence extending from the inner to the outer surface, which *superiorly* is roughened for attachment of the long inferior sesamoideal ligament, and below this presents a smooth surface coated with articular cartilage, lubricated with synovia, over which the flexor pedis perforans plays immediately after passing between the two terminal divisions of perforatus, which become attached to the extremities of this ridge inclined to the lateral surfaces of the bone. Between this ridge and the inferior articulatory surface the posterior part of the bone is unoccupied, merely presenting foramina for the passage of vessels into the bone. The *lateral surfaces* are exactly similar. Superiorly they are roughened for attachment of the broad lateral ligament of the pastern-joint. Inferiorly, inclined to the anterior surface, is a peculiar depression, which, with its fellow, seems as though the bone, when in a soft state, had been compressed between the thumb and finger. To this the lateral ligament of the pedal joint runs. The *anterior surface* of this bone is occupied by the expanded inferior part of extensor pedis tendon, which is attached somewhat firmly by areolar tissue.

OS PEDIS is an irregular bone, and since it receives no bony support from below is a **floating bone**. It is *symmetrical*, presenting two very similar halves, the inner of which is the smallest. At the extreme antero-inferior part may be seen in many cases a notch, a trace of the division, such as we see in ruminants, of this bone into two similar parts. It presents three surfaces: supero-posterior, infero-posterior, and antero-lateral. The **supero-posterior surface** centrally is articulatory, presenting a surface covered with cartilage, divided primarily into two parts. The *anterior part* is roughly heart-shaped, concave in an antero-posterior direction, elongated and slightly convex from side to side. Anteriorly it terminates in an obtuse point, extending in an upward direction on the posterior surface of the coronal process; posteriorly it presents an oblique angled indentation, into which the *posterior part* fits. The sides of the oblique angle meet the external convex margins of the articulatory surface at an acute angle outwardly and posteriorly. This surface centrally presents an antero-posterior broad ridge with a shallow concavity on either side, the inner being slightly the largest. It articulates anteriorly with os coronæ, and this portion is directly continuous with that between its posterior margins, with which os naviculare comes in apposition, and which is bounded posteriorly by a line extending from one posterior angle of the heart-shaped surface to the other. This surface of the bone is completed by a *roughened ridge on each side* running at first outwards and then directly backwards for about three eighths of an inch, from the posterior angles of the articulatory surfaces. The point where the ridges bend backwards form posteriorly smooth grooves, through which the plantar arteries run downwards in their course between the inferior broad ligament of the navicular bone and the perforans tendon to the foramen on the infero-posterior surface of os pedis.

The **postero-inferior surface** of the bone is mainly occupied by the **plantar surface**, semilunar in form, bounded by two margins, which meet posteriorly at acute angles. Thus this surface is broadest centrally, and gradually decreases in size posteriorly. It presents a slant from the greater or outer to the lesser or inner curved margin, and is perforated by foramina, which are most numerous at the angles and at the outer circumferent margin; and at the line of junction of this with the antero-lateral surface is a series of openings larger than the rest. Extending backwards from the posterior margin is a *roughened space*, bounded posteriorly by the articulatory surface for os naviculare, which on either side extends slightly on to this face of the bone. This roughened space is divided on either side into two parts by a groove extending from behind forwards, to terminate in a foramen. Through this the terminal portion of the plantar artery runs, and thus gains *a canal extending through the substance of the bone* from one foramen to the other, in which it anastomoses with its fellow, completing the **circulus anteriosus of the foot**. The walls of the grooves, as well as those of the canal, are perforated by foramina, through which branches pass outwards. The space between these grooves and the navicular articulatory surface is roughened for attachment of the inferior broad navicular ligament; it is perforated by articular foramina. That space between the grooves and the posterior margin of the plantar surface is about three eighths of an inch broad in the centre, and is rough for attachment of the extensor pedis tendon. The **antero-lateral surface** is largest centrally, and gradually diminishes in size posteriorly. At its superior part it presents a smoother portion of bone than the rest, of a pyramidal shape, having its apex superiorly placed. This is the **coronal process**; around its superior part, at about a quarter of an inch from its superior margin, runs a line marking out the inferior attachment of extensor pedis, which thus intervenes between the coronary secreting substance and the bone at this part. This surface is perforated by numerous foramina, and presents small irregular plates running from above, most marked near the inferior margin, where, also, at varying distances from each other, are large foramina for the passage of the anterior branches of the interosseous plantar artery, which assist to form the circumflex artery of the toe. Supero-laterally is a deep depression, smooth at its bottom for attachment of the lateral ligament of the coffin-joint. Behind this is the *ala*, running backwards for attachment of the lateral cartilage, which also seems to continue the bone from the posterior margin of this surface, assisting to form a foramen through which the lateral laminal branch of the plantar artery runs, coursing its way through a groove in a forward direction between the bone and the sensitive laminæ; at the anterior extremity of the bone it enters its substance through a large foramen. It sends off numerous branches, especially one running straight downwards, for which there is a special secondary groove, and which forms the commencement of the circumflex artery of the toe. The posterior margin of this surface presents superiorly the ala, inferiorly the angles of the plantar surface; between these two is the lateral laminal groove. They are rough, and are directly continuous with the lateral cartilage. In consequence of its porous nature, and the close reticulation of blood-vessels in its substance, os pedis seems to be soft, but its structure in the dried bone will be found to be remarkably *hard*.

Attached to the upper margin of the alæ of os pedis, extending superiorly for about one and a half inches, posteriorly for about one and a half inches behind the posterior extremities of the alæ, and extending farthest forward at the superior part of their anterior margin, are the **lateral cartilages of the foot**. These are peculiar shaped portions of cartilage which present two surfaces, three margins. The *external surface* is convex, extending at first upwards from the wing of os pedis with an outward inclination, which is increased along its posterior margin. Its anterior extremity becomes ultimately blended with the dense fibrous tissue which, in this situation, seems to closely unite the lateral ligament of the pedal-joint with the extensor pedis tendon at its inferior extremity. Posteriorly this surface terminates in a point at about one and a half inches from the ala of os pedis, for here the cartilage is inflected in a forward direction, forming that portion of the cartilage which, running along the inner surface of the wing of os pedis, gradually terminates in blending with the white fibrous tissue joining the expanded perforans tendon. Inwardly it is continued as a firm but fine layer of fibrous tissue, which serves to separate the fibrous frog from the tendon. The *inferior margin* of the cartilage in becoming attached to the bone helps to form the foramen through which the lateral laminal artery passes, and posteriorly, on its under surface, presents a triangular continuation of the plantar surface of the bone which affords attachment to the sensitive bars. The *internal surface*, as a whole, is concave, and looks in a direction inwards and slightly upwards. Its inferior part is formed by the upper surface of the *inflections of the cartilages*, which here are closely united to the fibrous frog, many of the elastic fibres of which become attached to it. This organ is separated from the main portion of this surface by a plexus of veins, which communicates with a corresponding plexus on the external surface, not only by branches proceeding through numerous foramina in the cartilage, but also by junction of their resulting vessels above the cartilage. The *superior margin* of the cartilage is about its thinnest part; the thickest is that just behind os pedis. In contact with the plexus of vessels on the external surface, centrally, lies the coronary secreting substance, superiorly skin, inferiorly the upper portions of the lateral laminæ, which are supplied by a special branch detached from the lateral laminal artery running backwards.

OS NAVICULARE (SHUTTLE BONE) is a *sesamoid* bone situated behind and slightly resting upon os pedis. It is elongated from side to side, and presents two surfaces, two margins, and two extremities. The *superior margin* looks upwards and backwards; it is straight, and presents some articular foramina. To it is attached the superior broad navicular ligament, which runs obliquely upwards and backwards to blend with the *perforans* tendon, and with the superior band of the stellate ligament. At its posterior edge a continuation of the posterior smooth surface of the bone is visible. This *posterior surface* is covered with fibro-cartilage, forming the anterior boundary of the navicular joint, and affording the perforans tendon a surface to play over. Its superior margin is straight, its inferior convex. It presents two level surfaces sloping slightly towards the centre, and thus rendering more prominent a ridge which here extends from above downwards across the surface. The *inferior margin* of the bone is convex, and posteriorly is rough, with numerous foramina, affording attachment to the inferior broad ligament (running from this to the postero-inferior surface of os pedis); anteriorly it presents a smooth articulatory surface, elongated from side to side, which moves arthrodially on a corresponding surface of os pedis, and is continuous by a smooth rounded edge with the *upper or anterior surface* of the bone. This is covered with articular cartilage, articulating with os coronæ; it has two elongated concave surfaces, which centrally are separated by a broad convexity. Its superior margin is straight, its inferior convex. The surfaces and margins culminate in the *two extremities*, which are angles, with a slight inclination in a posterior direction. To them are attached the stellate navicular ligaments, each of which, attached here by a single head, divides into three parts, one of which runs to the inner part of the wing of os pedis, blending with the inferior broad ligament, another runs to inner surface of the lateral cartilage, while the third, after uniting with the terminal expansion of the perforans tendon, passes upwards and divides, one part, after becoming attached to the inner surface of the superior part of the lateral cartilage being continued on to blend with the lateral ligament of the pastern-joint, while the other runs direct to the middle of the lateral surface of os suffraginis, where it is attached.

We have now examined the articulatory surfaces which enter into the formation of the coffin and navicular joints. The synovial membrane of the former covers the inner surface of part of extensor pedis tendon and of the lateral ligaments, and covers the anterior surface of both the broad navicular ligaments. We have here a peculiar, but not exceptional, union of a ginglymoid with an arthrodial joint. The arrangement of the tendinous structure around prevents the necessity for a capsular ligament. The synovial membrane of the navicular joint (or more correctly bursa), after covering the antero-superior part of the inferior portion of the tendon on reaching the superior broad ligament above, is reflected over its posterior surface as far as the bone, while from below it is reflected upwards over the posterior surface of the inferior broad ligament centrally, and laterally comes in contact with the stellate ligament.

The *blood-vessels and nerves* of the foot have been already noticed, but we may here repeat that each **metacarpal or metatarsal nerve** just above the fetlock breaks up into three divisions, which pass downwards over the lateral surface of the joint, in company with the plantar vein and artery. The *anterior division* runs forwards to supply the structures in front of the joint.

The **middle division or anterior plantar nerve**, crossing the artery from behind forwards, runs to the coronet, during the first part of its course situated behind the artery and vein.

The **posterior division or posterior plantar nerve** is situated behind the artery, and accompanies it to the posterior part of the foot, sending off fibres to accompany many of its branches; this nerve is divided in the "*low*" operation of neurotomy. The inferior extremity of the metacarpal nerve before division is involved in the "*high*" operation.

The flexor tendons must now be examined.

While passing through the sesamoid groove at the back of the fetlock the **perforatus tendon** extends round the perforans, and thus forms a sheath for it, which is deficient anteriorly, so that the theca between the tendons communicates with the sesamoid bursa. From the superolateral angles of os suffraginis portions of white fibrous tissue run to the posterior surface of the perforatus tendon, which shortly after bifurcates one of its divisions running to each extremity of the posterior fibro-cartilaginous surface of os coronæ. Between the divisions the **perforans tendon** runs, playing over the fibro-cartilaginous surface; to its posterior surface run white fibrous tissue layers from the infero-lateral parts of os suffraginis; thus reinforced it receives some white and yellow elastic fibres from the superior margin of the navicular bone (superior broad ligament) which run to its anterior surface. It then plays over the inferior surface of os naviculare, through the medium of the navicular bursa, and becomes inserted by a wide attachment to the posterior margin of the plantar surface of os pedis. Its under surface is separated from the fibrous frog by a white fibrous layer, which connects the lateral cartilages at their inferior inflections.

ARTERIAL TREE.

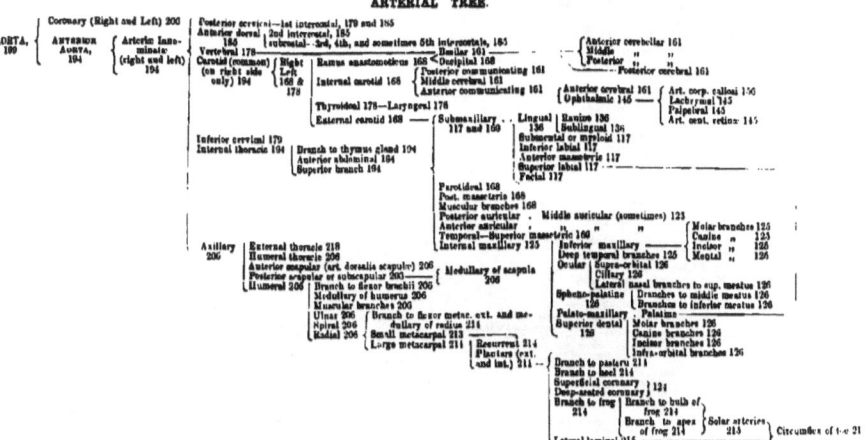

ARTERIAL TREE—continued.

```
Bronchial and œsophageal 192
Intercostals (9th to last)          Medullary of dorsal vertebræ 186
                186                 Branch to spinal cord and its membranes 187
                                    Dorsal branches 187
                                    Cutaneous, muscular, and pleural branches 187
Phrenic or diaphragmatic 199
        Superior branches 238
Lumbar  Inferior branches 238
  238   Medullary of lumbar vertebræ 238
        Branches to spinal cord and its membranes 238
                Gastric   Superior 237
Cœliac axis     237       Inferior 237
     237        Splenic 237 ,  Left gastric 237
                Hepatic 229    Branches to the liver 229
                               Duodenal branch 229   Right gastric 229
                               Pancreatic 237         Intestinal 229
Anterior mesenteric     Duodenal 237
     237                Mesenteric (15—18 in number) 237
                        Cæcal 237
                        Colic 237    Two branches to double colon 237
                                     Branch to single colon 237
Posterior mesenteric    Branch to single colon 237
     237                Branch to rectum 237
Renal   Branch to prerenal capsule (sometimes from post. aorta) 237
 237    Branches to kidney 237
                        Superior testicular 237
Spermatic (male) 238    Inferior testicular 237
Ovarian (female) 238    Branches to spermatic cord 237
                        Branches to epididymis 237
Artery of cord (male) 263   Sometimes from external iliac
Uterine (female) 263
Middle sacral (sometimes) 265 — Middle coccygeal (sometimes) 265
                                    Umbilical    Round ligament of the bladder 264
Internal iliac   Artery of the bulb              Vesical branches 264
     264            264                Vesical branches 264
                                       Anal and perineal branches 264
                                       Branches to bulb of penis 264
                    Arteria innominata 264
         Obturator   Pubic 264
            264      Ischiatic 264
                     Internal pudic — Posterior dorsal artery of the penis 264
         Gluteal 263
         Lateral sacral 265 — Lateral (sometimes middle) coccygeal 265
External iliac   Circumflex artery of the ilium 263
     262         Arteria profunda   Epigastric   Post. abdominal branches 263
                 femoris 263          263         External pudic   Anterior dorsal artery
                                                      263           of penis 263
```

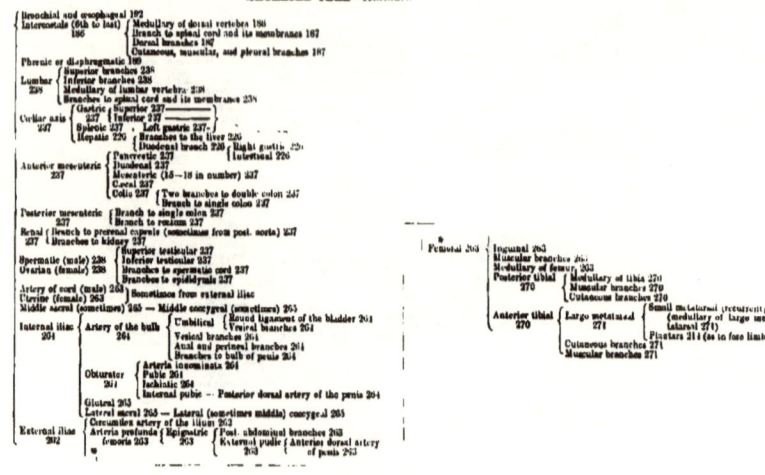

```
                     Femoral 263    Inguinal 263
                                    Muscular branches 269
                                    Medullary of femur 263
                                    Posterior tibial    Medullary of tibia 270
                                         270            Muscular branches 270
                                                        Cutaneous branches 270
                                    Anterior tibial     Large metatarsal   Small metatarsal (recurrent)
                                         270                 271            (medullary of large me-
                                                                            tatarsal 271)
                                                         Cutaneous branches 271    Plantars 271 (as in fore limb)
                                                         Muscular branches 271
```

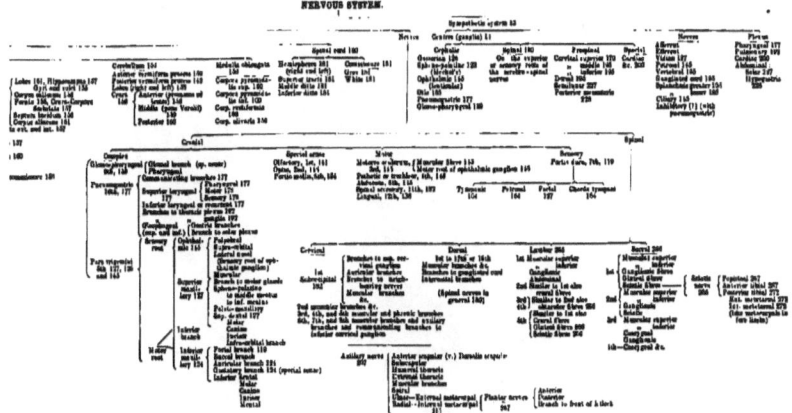

INDEX.

	PAGE
Abdomen	216
— divisions of	231
— lining membrane of	222
— muscles of	220
— regions of	231
— viscera of	224
Abdominal rings	219
Absorbent glands, see Glands, lymphatic.	
Absorbents, see Lymphatics.	
Acetabulum	82
Acini	226
Afferent lymphatics	19
— vessel of kidney	240
Age of subject for dissection	3
— as shown by teeth	131
Air cells	191
— sacs	192
Alæ of atlas	48
— nostrils	139
— thyroid cartilage	173
Albuginea testis, tunica	246
Allantois	255
Allantoid sac	255
Alveoli	34
Amnion	255
Amphiarthrosis	100
Ampullæ	163
Anastomosis (of arteries)	21
Anatomy	1
— comparative	1
— descriptive	2
— general	1
— pathological	2
— philosophical	2
— transcendental	1

	PAGE
Angle of inferior maxilla	41
— mouth	120
— palpebral fissure	120
— rib	58
Ankle	92
Annular cartilage	122
— ligaments	11
Annulus ovalis	197
Anthropotomy	1
Antrum	142
Anus	235
Aorta	199
— anterior	194
— posterior	236
— valves of	199
Apophysis	8
Apparatus, lachrymal	121
Appendix	197
Aqueduct of Sylvius	158
Aqueous humour	149
Arachnoid (cranial)	154
— fluid	155
— (spinal)	180
Arch, crural	219
— hæmal	42
— fibular	92
— ischial	85
— neural	42
— palatine	38
— radio-ulnar	69
— vertebral	43
— zygomatic, see Zygomatic ridge.	
Arcus senilis	147
Arteries, general description of	20
— injection of	2

290 INDEX.

	PAGE
Arteries, coats of	21
— descriptive anatomy of:	
— abdominal	221
— aorta	199
— — anterior	194
— — posterior	236
— auricular	123
— axillary	206
— basilar	161
— bronchial	192
— of bulb	264
— carotid, common,	168, 178, 194
— — external	168
— — internal	168
— centralis retinæ	145
— cerebellar	161
— cerebral	161
— cervical, inferior	179
— — posterior	179, 185
— ciliary	145
— circulus arteriosus pedis	214
— circumflex, of ilium	262
— toe	215
— coccygeal	265
— cœliac	237
— cœcal	237
— colic	237
— communicating	161
— of cord	263
— coronary, of heart	200
— — foot	214
— corporis callosi	156
— dental, superior	126
— dorsal	185
— dorsalis scapulæ	217
— diaphragmatic, see Phrenic.	
— duodenal of hepatic	226
— emulgent	237
— epigastric	263
— facial	117
— femoral	263
— of frog	214
— gastric, superior	231
— — inferior	231
— — right	231
— — left	231
— gluteal	265
— hepatic	226
— humeral	206
— — thoracic	206
— iliac, external	262

	PAGE
Arteries, iliac, internal	264
— infra-orbital, see Superior dental.	
— inguinal	263
— innominate, right	194
— — left	194
— — branch of obturator	264
— intercostals	185, 186
— — first	179
— ischiatic	264
— labial, superior	117
— lachrymal	145
— laminal, lateral	214
— laryngeal	178
— lingual	136
— lumbar	238
— mammary	263
— masseteric	117
— maxilary, internal	125
— — inferior	125
— mesenteric	227
— — anterior	237
— — posterior	237
— metacarpal, small	213
— — large	214
— metatarsal, large	271
— — recurrent	271
— nasal, lateral	126
— obturator	264
— occipital	168
— œsophageal	192
— ophthalmic	145
— orbital	145
— ovarian	238
— palato-maxillary	126
— pectoral, external, see Thoracic.	
— phrenic	189
— plantar	214
— profunda femoris	263
— pubic	264
— pudic, external	263
— — internal	264
— pulmonary	198
— radial	206
— ramus anastomoticus	168
— ranine	136
— recurrent	270
— renal	237
— sacral, middle	265
— — lateral	265

INDEX. 291

	PAGE
Arteries, scapular, anterior	206, 217
— — — posterior	206, 217
— solar	286
— spermatic	238
— spheno-palatine	126
— spiral	200
— splenic	237
— of stomach	236
— sublingual	136
— submaxillary	117, 169
— subscapular	217
— subzygomatic, see Temporal.	
— supra-orbital	126
— temporal	169
— — deep	125
— thoracic, external	218
— — internal	194
— thyroid	178
— tibial, anterior	270
— — posterior	270
— ulnar	206
— umbilical	264
— uterine	238, 263
— utero-ovarian, see Ovarian.	
— vertebral	178
— vesico-prostatic	264
Arthrodial joint	100
Arthrology	199
Articular cartilage	17
— eminences	9
— laminæ	10
Articulation, see Joint.	
Arytenoid cartilages	172
Ass	3
Auditory canal, external	123
— meatus, external	123
— — internal	32
Auricle of heart, right	197
— left	199
Auriculo-ventricular furrows	196
— — opening, right	198
— — — left	199
— — valves	199
Axonoidal joints	101
Azygos vena	187
Bacillary layer	148
Back	183
Barb	132

	PAGE
Bars of hoof	276
Base of mouth	128
— brain	160
Basement membrane	14
"Beastlings"	232
Belly of muscle	12
Bicuspid valve	199
Bile duct	233
Bladder, gall, absence of	225
— urinary	241
— — fœtal	255
— — ligaments	241
Blind spot	148
Blood, circulation of.	
— vessels	19
Body of thyroid	166
Bone, general anatomy of	7
— cavities of	10
— composition of	6
— classes of	8
— tissue, kinds of	8
— eminences of	10
— formation and growth of	7
— Haversian canals	8
— — systems	8
— lacunæ of	8
— lamellæ	8
— medulla of	8
— periosteum of	9
— processes of	10
— organization of	9
— long, round	8
— flat	9
— short	9
— irregular	9
— astragalus	93
— atlas	47
— axis	49
— calcis	92
— carpal	72
— coccygeal vertebræ	56
— coronal	281
— cranial	150
— cuboid	96
— cuneiform of carpus	73
— — magnum (tarsus)	94
— — medium —	95
— — parvum —	96
— ear	164
— ethmoid	29
— face	33

INDEX.

	PAGE		PAGE
Bone, femur	85	Bone, suffraginis	79
— fibula	91	— tarsal	92
— foot	281	— temporal, petrous	32
— forearm	68	— — squamous	31
— fore extremity	61	— tibia	89
— frontal	27	— trapezium	74
— hind extremity	82	— trapezoid	75
— hock	92	— triquatral	26
— humerus	65	— turbinated, superior	33, 140
— hyoid	133	— — inferior	33, 140
— ilium	82	— ulna	70
— incus	164	— unciform	76
— innominate	82	— vertebral, cervical	46
— ischium	84	— — coccygeal	56
— lachrymal	36	— — dorsal	50
— lunare	72	— — false	44
— magnum	75	— — lumbal	52
— malar	38	— — sacral	54
— malleus	165	— — true	44
— maxillary, anterior	33	— — typical	42
— — inferior	40	— vomer	39
— — superior	34	— zygomatic	38
— metacarpal, large	77	Brain	155
— — small	78	—— circle of Willis	161
— metatarsal, large	98	—— ventricles	158
— — small inner	98	Bronchi	191
— — — outer	98	Brunner's glands	233
— nasal	33	Buccal salivary glands	118
— navicular	285	Bulb, olfactory	160
— occipital	24	— penis	244
— palatine	37	— vagina	257
— parietal	26	Bursa	15
— patella	88		
— pedal (fore)	282	Cæcum	234
— — (hind)	98	— caput coli	234
— pelvic	82	Calamus scriptorius	159
— pisiform	77	Calcaneo-cuboid ligament	114
— pterygoid	38	Calyces of kidney	240
— pubic	85	Canals	10
— quadrate	101	Canal, alimentary, *see* Stomach, &c.	
— radius	68		
— ribs	56	— Haversian	8
— sacral	54	— auditory external	123
— scaphoid	72	— of Fontana	148
— scapula	61	— of Petit	147
— sesamoid	80, 98	— inguinal	219
— shuttle	285	— spinal	180
— skull	24	Canaliculi	8
— sphenoid	30	Cancellated structure	8
— stapes	164	Canine teeth	132
— sternal	60	Canthi of eyelids	121

INDEX. 293

	PAGE		PAGE
Capillaries	26	Cauda equina	266
Capsular ligaments	100	Caudatus lobus	225
Capsule, Glisson's	226	Caul	230
— of lens	149	Cavernosum, corpus	243
— pre-renal	236	Cavernous sinus	154
— supra-renal, *see* Pre-renal	236	Cavities of bones	9, 10
— renal	239	— heart	196
Cáput gallinaginis	242	— reserve	130
Cardia of stomach	229	Cavity, abdominal	230
Cardiac ligament	224	— alveolar	33
Cariniform cartilage	60	— cranial	150
Carneæ columnæ	198	— glenoid	63
Carpus, bones	71	— pulp	132
— articulations	107	Cells, ethmoid	30
Cartilage	7	— hepatic	226
General:		— mastoid	164
— cellular	8	— nerve	14
— elastic	7	Cementum	129
— fibro-	7	Centralis retinæ arteria	145
— hyaline	7	Centrum ovale major	156
— permanent	7	— — minor	156
— temporary	8	— vertebral	43
— true	7	— nerve	13
— articular	7	Cerebellum	159
Particular:		Cerebro-spinal nervous system	13
— annular	122		
— arytenoid	175	Cerebrum	155
— auricular	122	— crura of	160
— cariniform	60	Ceruminous glands	123
— costal	58	Cervix vesicæ, *see* Neck of bladder.	
— cricoid	174		
— elongation of scapula	62	— uteri, *see* Neck of uterus.	
— — ribs	58	Chambers of the eye	149
— — eusiform	60	— nasal	139
— epiglottis	174	Channels	10
— of foot (lateral)	284	Cheek	118
— inter-articular discs of stifle-joint	111	Chestnut	208
		Chorda tympani	124
— — temporo-maxillary joint	101	Chordæ tendineæ	198
		Chorion	253
— larynx	173	Choroid	147
— nictitans	143	— plexus of lateral ventricles	155
— of nostrils	139		
— septum nasi	173	— — of third ventricle	155
— scutiform	122	Ciliæ	16
— tarsal	120	Ciliary arteries	16
— thyroid	173	— ligament	148
— of trachea	176	— processes	147
— xiphoid	60	— zone	149
Curuncula lachrymalis	121	Ciliated epithelium	16
Carunculæ myrtiformes	251	Cineritious brain matter	156

INDEX.

	PAGE		PAGE
Circle of Willis	161	Cornua of os hyoides, long	135
Circulation, blood, fœtal	253	— of spinal cord	181
— through liver	226	— of thyroid cartilage	173
Circulus arteriosus pedis	214	— of ventricles (lateral)	156
Circumferential cartilage (hip)	111	Corona glandis	244
Circumflex artery of the toe	215	— tubulorum	234
Clitoris	251	Coronal suture	102
Coats of stomach	230	Coronary arteries	200
Coccyx	56	— fibrous band	280
Coccygeal vertebræ	56	— frog band	275
Cochlea	163	— secreting substance	278
Cœliac axis	237	Corpus albicans	161
Colon, double	234	— callosum	156
— single	235	— cavernosum	243
Colostrum	253	— dentatum	159
Colour of the hoof	276	— geniculatum	157
Column, vertebral	42	— luteum	248
Columnæ carneæ	198	— olivare	159
Columns of spinal cord	181	— spongiosum	244
Commissures of hoof	274	Corpora arantii	195
— grey	157	— striatum	157
— soft	157	— pyramidalia	160
— of spinal cord	181	— restiformia	160
— anterior	158	— nigra	148
— posterior	158	— quadrigemina	158
Communicating arteries	161	Corpuscles, Malpighian, of	
— foramina, anterior	158	kidney	240
— — posterior	158	— — spleen	236
Compact tissue of bone	8	Costæ	26
Compound glands	19	Costal cartilages	58
Conchial cartilage	122	Cotyloid cartilage	111
Condyles	10	— cavity	10
— of femur	88	Cowper's glands	242
— of humerus	67	Cranium	24
— of inferior maxilla	40	Cranial nerves	160
— of occiput	24, 25	Cremasteric fascia	245
Conjunctiva	121	Crest	10
Constrictors of the pharynx	138	— of occiput	26
Contorti tubuli	240	— of ilium	83
Convolutions of the brain	155	Cribriform plates of the	
Coracoid process	61	ethmoid	30
Cord, artery of	238	Cricoid cartilage	174
— spermatic	245	Crista galli process	29
— spinal	180	Crown of tooth	128
— umbilical	255	Crucial ligament	112
Cordæ vocales	175	Crura cerebri	160
— Willisii	153	— cerebelli	159
Corium	16	— of corpus cavernosum	243
Cornea	46	— of diaphragm	189
Corneal cartilage	139	— fornix	156
Cornua of os hyoides, short	134	— of pineal body	158

INDEX. 295

	PAGE		PAGE
Crura of sphenoid bone	31	Ducts, nasal	143
Crural arch	219	— pancreatic	224
Crusta petrosa	129	— of Stenson	141
Crypts, mucous	242	— thoracic	195
Crystalline lens	148	— Wharton's	132
Curvatures of stomach	229	— ejaculatory	247
— spleen	236	Ductus ad nasum	143
Cuticle, see Epidermis.		— arteriosus	198, 254
Cuticular layer of stomach	230	— venosus	253
Cutting the teeth	130	Ductless glands	19
Cylinder, axis	14	Duodenum	233
		Dura mater of the brain	153
Dark chamber	149	— — cord	180
Decidua membrana	250		
Decussation, optic	160	Ear	163
— of motor fibres	160	— external	122
Dentata	49	— internal	163
Dentition (table)	131	— middle	164
Dentine	129	— wax, see Cerumen.	
Depressions on bones, see Cavities.		Ejaculatory ducts	247
		Elastic tissue	7
Dermis	17	Elastica, cornea	146
Descent of testis	246	Elongation, cartilage, of rib	58
Destruction of subject	4	— — of scapula	62
Detrusor urinæ	242	Eminences of bones	9
Development of teeth	130	Enamel	129
Diaphragm	188	Encephalon	162
Diapophysis	8	Endocardium	196
Diarthrosis	100	Endolymph	163
Digital impressions	10	Ensiform cartilage	60
Digastric muscles	12	Epidermis	434
Digestive organs, see Stomach, &c.		Epididymis	247
		Epiglottis	289
Dilators of the pharynx	138	Epiphysis	8
Diploë	9, 150	Epithelium, ciliated	16
Disc, optic	148	— columnar	16
Dissection of head	115	— spheroidal	16
— of neck	166	— stratified	16
— of abdomen	216	— tesselated	15
— of back	183	Erectile tissue	236
— of fore extremity	201	Ethmoid bone	29
— of hind extremity	257	— cells	30
Division of the subject	4	— fossæ	151
Dorsal vertebræ	50	— sinuses	29, 142
Dorsum scapulæ	63	Eustachian tube	137
— ilii	83	Extremities, anterior, see Fore limb.	
Drum of the ear, see Tympanum.		— posterior, see Hind limb.	
Ducts, bile	225	Eye	146
— of glands	19	— muscles of	144
— hepatic	225	Eyeball	146

INDEX.

	PAGE
Eyelids	120
Eyelid, third	143
Eyelashes	121
Excrescence, cauliflower	250
External ear	122
Extremity, dissection of fore	201
— — of hind	257
Face	24
— bones of	33
Facial nerve	119
Falciform ligament of bladder	241
— — of liver	22
Fallopian tube	249
False nostril	116
— ribs	58
— tongues	254
— vertebræ	44
Falx cerebri	153
Fang of tooth	128
Faschia or fascia	11
— of arm	210
— abdominalis superf.	216
— cremasteric	245
— intercolumnar	245
— lata	258
— lumbar	183
— psoas	236
— of quarter	257
Fasciculi of muscle	12
Fat in orbit	143
Fatty mane	166
Fauces, isthmus of	137
— pillars of	137
Feelers	120
Female generative organs	273
Femoral space	260
Fenestra ovalis	164
— rotunda	164
Fenestrated membrane of Henle	21
Fibres of Corti	164
— nerve	14
— muscle	12
Fibrillæ	12
Fibro-cartilage	7
Fibrous coronary band	280
Fibrous tissue	7
Fibular arch	92
Filiform papillæ	133
Filum terminale	180

	PAGE
Fimbriated extremity	249
Fissure	10
Fissures of spinal cord	181
— cerebrum	155
— palpebral	121
— of Sylvius	161
— of liver, transverse	225
Flat bones	9
Fixed joints	99
Fleur épanouie	250
Flexure, sigmoid, of double colon	234
Foal, peculiarities of	255
Fœtal appendages	253
— circulation	253
Follicular glands	18
Follicles of Lieberkühn	232
Fontana, canal of	148
Foot	273
— bones of	281
— nerves of	287
— vessels of	280
— tendons of	287
Foramen auditorium externum	32, 152
— — internum	32, 151
— communicating anterior	157
— — posterior	158
— condyloideum	25, 151
— dextrum	188
— incisorum	34
— infra-orbitale	35
— lacerum basis cranii	151
— — orbitale	30, 152
— lachrymal	37
— magnum occiputum	151
— — vertebral	49
— maxillary, anterior	41
— — posterior	41
— medullary	9
— menti	41
— of Monro	157
— obturator	84
— optic	152
— orbitale internum	152
— ovale	196, 254
— palato-maxillary	38
— pathetic	152
— rotundum	30, 152
— sinistrum	189
— Soëmering	148

	PAGE		PAGE
Foramen, spheno-palatine	37	Ganglion, inferior	195
— styloid	32, 152	— lenticular	145
— subsacral	56	— middle	195
— subsphenoidal	30	— Meckel's	123
— supersacral	56	— ophthalmic	145
— supra-orbital	28	— otic	165
— of Winslow	223	— semilunar	227
Foramina	10	— post-mesenteric	228
— articular	9	Ganglionic nervous system	13
— of cranium	151	Gaps, intervertebral	43
— intervertebral	43	Gastric glands	230
— Thebesii	197	Gastro-hepatic omentum	223
Fore limb, articulations of	106	— phrenic omentum	224
— bones of	61	— splenic omentum	224
— dissection of	201	— colic omentum	223
Foreskin	244	Generative organs, female	248
Fornix	381	— — male	243
Fossa	10	Geniculata, corpora	157
— antea spinatus	63	Genu of corpus callosum	156
— ethmoid	150	Ginglymoid joint	101
— intercondyloid (femur)	88	Glands	18
— navicularis, male	243	— Brunner's	233
— — female	251	— buccal	118
— olfactory	30	— ceruminous	123
— optic	29	— Cowper's	242
— ovalis	197	— ductless	19
— postea spinatus	63	— gastric	230
— renal	239	— labial	119
— temporal	125	— lachrymal	143
— trochanteric	86	— of large intestine	233
Frœnum linguæ	133	— lymphatic	23
Fræna (synovial)	15	— — inguinal	260
Freely moveable joints	100	— — popliteal	272
Fringes, synovial	15	— — mesenteric	227
Frog, horny	276	— lymphatic axillary	206
— sensitive	280	— — bronchial	192
— fibrous or fatty	280	— — prescapular	201
— stay	276	— parotid	137
Frontal sinus	142	— mammary	152
Fundus of bladder	241	— Meibomian	120
Fungiform papillæ	133	— mucous, of stomach	230
Furrows	10	— molar, see Buccal.	
— of heart	196	— of Pacchioni	154
Galactophorous sinuses	252	— palatine	137
Gall duct, see Bile duct.		— pancreas	224
Ganglia	13	— parotid	167
— of heart	200	— peptic	230
Gangliated cord	195	— Peyer's	233
Ganglion, Arnold's, see Otic.		— prostate	243
— cervical, superior	178	— salivary	167
— Gasserian	126	— sebaceous	17

INDEX.

	PAGE
Glands, of skin	17
— of small intestine	232
— solitary	232
— sublingual salivary	133
— submaxillary salivary	167
— sudoriparous	17
— thymus	200
Glans penis	244
Glenoid cavity	10
Glisson's capsule	226
Globe of the eye	146
— epididymis	247
Glosso-pharyngeal nerve	139
Glottal opening	175
Glottidis, rima	175
Gomphosis	99
Graafian vesicles	248
Grey nerve-fibres	13
Gristle, see Cartilage.	
Groove	10
— primitive dental	130
Gubernaculum testis	246
Gullet, see Œsophagus.	
Gums	128
Gustatory nerves	136
Guttural pouches	137
Gyri	155
Gyrus fornicatus	155
Hæmal arch	42, 150
— spine	43
Hæmorrhoidal, see Rectal.	
Hair	17
— follicles	17
— root	17
— stem	17
— bulb	17
Hamular process of ulna	71
Harmonia	99
Haversian canals	8
— systems	8
Haw, see Cartilago nictitans.	
Head, dissection of	115
— bones of	24
Heads	10
Heart	192
— annulus ovalis	197
— arteries of	194
— auricles of right	197
— — left	199
— — appendices	199

	PAGE
Heart, cavities of	197
— chordæ tendineæ	198
— columnæ carneæ	198
— corpora arantii	199
— endocardium	196
— fibrous rings of	196
— fœtal	254
— foramina Thebesii	197
— fossa ovalis	197
— furrows	196
— lining membrane	196
— muscular fibres of	197
— musculi papillares	198
— — pectinati	197
— nerve supply	200
— orifice, auriculo-ventricular, right	197
— — — left	199
— — pulmonary	198
— — aortic	199
— sinus Valsalvæ	199
— systemic	196
— tuberculum Loweri	197
— valves, auriculo-ventricular, right	197
— — — left	199
— — Eustachian	197
— — mitral or bicuspid	199
— — pulmonary	196
— — semilunar aortic	199
— — — pulmonary	198
— — systemic	196
— — Thebesii	197
— ventricle, right	198
— — left	199
Heart's lymph	23
Hemispheres of brain	155
— cord	181
Hepatic cells	226
— duct	225
Hiatus	10
— aorticus	188
— sphenoideal or orbital	152
Hilum of kidney	239
— spleen	236
Hind limbs	257
— joints of	111
Hip-joint	10
Hippocampus	157
Hippotomy	1
Histology	1

INDEX. 299

	PAGE		PAGE
Hock-bones	92	Intestines, large, sigmoid flexure	234
— joint	113		
Hoof	273	— — mucous membrane of	230
— hind	278	— — muscular coat of	232
— commissures	274	— — rectum	235
— coronary frog-band	275	— — serous coat of	231
— frog of	277	— small, duodenum	233
— frog-stay of	277	— — glands of	232
— heels of	274	— — ileum	233
— horn of	274	— — jejunum	233
— laminæ of	275	— — mucous membrane of	232
— quarters of	274	— — muscular coat of	232
— sole of	276	— — serous coat of	231
— toe of	274	Intralobular veins	226
— wall of	274	Involuntary muscle	11
Horn	271	Iris	148
— secreting layer	80	Irregular bones	9
Horsehair	18	Ischial arch	85
Humour, aqueous	149	Ischium	84
— third	149	Isthmus of the fauces	158
— vitreous	149	Iter a tertio ad quartum ventriculum	158
Hymen	251		
Hypochondriac regions	231	Iter e cerebello ad testis	158
Hypogastric plexus	227		
— region	231	Jacobi, tunica	148
Hypoglossal nerve	136	Jacobson's canal	141
		Jaw, lower	24
Ileo-cæcal valve	233	— upper	24
Ileum	233	Jejunum	233
Iliac artery, external	238	Joints, amphiarthrotic	100
— — internal	238	— arthrodial	100
Ilium	82	— atlaxoid	102
Imprint	10	— atloido-occipital	102
Incisions (primary, through skin)	4	— axonoides	101
		— carpal	107
Incisors	129	— carpo-metacarpal	107
Incisive opening	34	— coffin	109
Incus	164	— costal	105
Inferior maxillary sinus	142	— diarthrodial	100
Infundibulum	161	— elbow	106
Infundibuli	129	— fetlock	108
Inguinal canal	219	— fixed	99
Injection of vessels	2	— of fore extremity	106
Inosculation	21	— ginglymus	100
Instruments	2	— harmonia	90
Interlobular veins	226	— of hind extremity	111
Intestines	231	— hip	111
— large	234	— hock	113
— — cæcum	234	— hyoid	134
— — colon, double	234	— knee	107
— — — single	235	— moveable	99

INDEX.

	PAGE		PAGE
Joints, partially moveable	100	Larynx, cartilages of	174
— pastern	109	— — arytenoid	175
— pedal	109	— — cricoid	174
— radio-carpal	107	— — epiglottis	174
— — ulnar	106	— — thyroid	173
— shoulder	106	— ligaments	175
— schindylesis	99	— mucous membrane of	176
— stifle	111	— muscles of	176
— sutural	99	— nerves of	177
— synarthrodial	99	— ventricles of	176
— tarsal	113	— vocal cords of	175
— tarso-metatarsal	113	Lateral cartilage of foot	284
— vertebral	103	Leg, see Limb.	
		Lens, crystalline	149
		— capsule of	149
Kidney	239	Lenticular ganglion	145
— adipose tunic of	239	Lieberkühn's follicles	232
— calyces	240	Ligaments	119
— capsule	239	— annular	11
— corpuscles of Malpighi	240	— of bladder	222
— cortical substance	240	— calcaneo-cuboid	114
— hilum	239	— capsular	11
— left	239	— cardiac	219
— medullary substance	240	— of bladder	241
— pelvis of	240	— ciliary	148
— pyramids of Malpighi	240	— coraco-scapular	62
— right	239	— cotyloid	111
Knee	71	— crico-trachealis	176
— joint	107	— — thyroid	173
		— crucial, of stifle	112
		— denticulatum	180
Labia, see Lips.		— extensorum	212
— majora	251	— interspinous	104
— minora	251	— of liver	283
Labyrinth	163	— nuchæ	104
Lachrymal canals	143	— obturator	111
— foramen	37	— odontoid	103
— gland	143	— of patella	112
— puncta	143	— of the pelvis	109
— sac	143	— Poupart's	219
— tubercle	36	— pubis femoral	111
Lacunæ	8	— sacral	104
— of urethra	241	— stellate	105
Lacteals	23, 227	— — of navicular	286
Lambdoidal suture	102	— — of rib	105
Lamina cribrosa	146	— subcarpal	210
— spiralis	164	— superior sesamoideal	208
Laminæ of bone	8	— supra-spinous	104
— vertebræ	42	— suspensory, superior	210
— horny	275	— — penis	216, 245
— sensitive	279	— tarsi	120
Larynx	173		

INDEX.

	PAGE
Ligaments, triangular penis	84
Limb, see Extremities.	
— bones of fore	61
— — hind	82
— articulations of fore	106
— — hind	111
Linea alba	228
Lips	119
Liquor Morgagni	150
Liver, cells of	226
— circulation through	226
— excretory apparatus of	224
— Glisson's capsule of	226
— ligaments of	223
— lobes of	225
— lobulus caudatus	225
— — scissatus	225
— — vessels of	226
Lobes of cerebrum	161
— cerebellum	159
— liver	225
— lungs	191
Lobule, mastoid	161
Lobules of liver	225
— testis	246
Lobulus caudatus	225
— scissatus	225
Locus perforatus anticus	161
— — posticus	161
Longitudinal sinus	153
Long round bones	8
Linear bone	72
Lungs, air-cells of	191
— lobes of	191
— roots of	191
— vessels, pulmonary	196
— — bronchial	192
Lutea, macula	148
Lymph hearts	23
— spaces	23
Lymphatics	22
— glands, see Glands.	
— afferent	23
— efferent	23
Macula lutea	148
Male generative organs	243
Malleolus	91
Malleus	165
Malpighian corpuscles of kidney	240

	PAGE
Malpighian corpuscles of spleen	236
— pyramids	240
Mammary glands	252
— ducts of	252
— sinuses of	252
Marrow	8
Mastoid cells	32, 164
— lobule	161
Maxillary sinuses	142
Meatus, auditory, external	123
— — internal	243
— of nasal chambers	141
Mediastina	189
Mediastinum testis	286
Medulla spinalis, see Spinal cord.	
— oblongata	159
— ossium, see Marrow.	
Medullary canal	8
— foramen	9
— matter of brain	156
— — kidney	240
— sheath	14
Meibomian glands	120
Membrana decidua	250
— fusca	147
— nictitans	142
— pupillaris	149
— tympani	164
Membrane of Demeurs	149
— of Descemet	149
— hyaloid	149
— simple	14
— compound	15
— mucous	16
— serous	15
— Schneiderian	286
— synovial	15
— tentorial	153
— winking	143
Membranes of brain	153
— foetus	253
— spinal cord	180
Membranous labyrinth	163
— semicircular canals	163
— urethra	242
Meninges, see Membranes.	
Mesentery	226
Meso-cœcum	222
— colon	222

	PAGE		PAGE
Meso-cæcum, rectum	222	Muscles, buccinator	118
Milk	252	— caninus	118
— teeth	129	— caput magnum	203
Mitral valve	199	— — medium	203
Moderator bands	198	— — parvum	204
Modiolus	163	— cervical trapezius	183
Molar teeth	130	— complexus major	171
Monorchid	246	— — minor	171
Monro's foramen	157	— compressor coccygis	272
Morgagni, liquor	150	— constrictors of pharynx	138
Motor nerve-roots	181	— coraco-humeralis	204
Motores oculorum	145	— — radialis	202
Motorial plates	14	— cornealis transversus	140
Mouth	127	— costarum superficialis	104
Moveable joints	100	— — transversalis	104
Mucous membrane	16	— — levatores	186
— of intestines	232	— crico-arytenoideus posticus	176
— lachrymal apparatus	121	— — lateralis	176
— larynx	175	— — pharyngeus	138
— mouth	133	— — thyroideus	176
— nose	141	— curvator coccygis	272
— œsophagus	172	— depressor ani	235
— palate	138	— — coccygis	272
— pharynx	138	— — labii superioris	119
— stomach	229	— — oculi	144
— tongue	133	— detrusor urinæ	135, 242
— trachea	178	— diaphragm	188
— tympanum	165	— digastricus	135
— ureters	240	— dilator naris anterior	140
— urinary bladder	241	— — — inferior	119
— uterus	250	— — — superior	116
— vagina	251	— of external ear, extrinsic	123
— vulva	251	— — intrinsic	122
Muscles	11	— of internal ear	165
— voluntary	12	— erector coccygis, see Levator.	
— involuntary	12		
— varieties of	12	— extensor metacarpi magnus	
— abdominal	219		211
— abductor oculi	144	— — — obliquus	211
— — of ear	144	— — pedis	212
— accelerator urinæ	244	— — — (hind)	260
— adductor oculi	144	— — suffraginis	212
— — auris	123	— flexor brachii	202
— anconeus	205	— — metacarpi externus	213
— of anus	335	— — — medius	209
— antea spinatus	202	— — — internus	209
— attolentes aurum	126	— — metatarsi	269
— attrahentes aurum	126	— — pedis accessorius	269
— auricular	126	— — — perforans	210, 269
— biceps adductor femoris	261	— — — — hind	269
— — rotator tibialis	259	— — — perforatus	210

INDEX.

	PAGE
Muscles gastrocnemius externus	267
— — internus	268
— gemini	267
— genio-hyoideus	134
— — hyoglossus	134
— gluteus externus	257
— — internus	258
— — maximus	258
— — medius, see Maximus	258
— gracilis	260
— humeralis externus	205
— hyo-epiglottideus	174
— hyoglossus brevis	134
— — longus	135
— — parvus	135
— hyo-pharyngeus	138
— hyo-thyro-pharyngeus	138
— hyoideus magnus	135
— — parvus	135
— — transversus	134
— iliacus	261
— ilio-femoralis	262
— inferior oblique, of eye	144
— intercostalis externus	186
— — internus	186, 218
— intertransversalis colli	179
— — dorsi	186
— ischio-femoralis	267
— — tibialis	259
— — trochanterius	267
— of larynx	176
— lateralis sterni	218
— latissimus dorsi	183
— levator ani	235
— — coccygis	272
— — costæ	186
— — humeri	169, 201
— — labii superioris alæque nasi	115
— — palpebræ superioris externus	120
— — — internus	143
— — oculi	144
— longissimus dorsi	184
— longus colli	178
— lumbrici	109
— masseter externus	116
— — internus	123
— mouth, which close	125
— mylo-hyoideus	123

	PAGE
Muscles, nasalis longus labii superioris	116
— obliquus abdominis externus	220
— — — internus	220
— — capitis anticus	170
— — — inferior	170
— — — superior	170
— obturator externus	267
— — internus	267
— orbicularis oris	120
— — palpebrarum	120
— palato-pharyngeus	137
— panniculus	115, 166, 216
— parotides auricularis	166
— pectineus	261
— pectoralis anticus	217
— — magnus	217
— — transversus	217
— peroneus	269
— of pharynx	138
— plantaris	268
— popliteus	268
— postea spinatus	202
— psoas magnus	255
— — parvus	256
— peterygoideus	124
— pyriformis	267
— radialis accessorius	210
— rectus abdominis	220
— — capitis anticus major	179
— — — — minor	179
— — — posticus major	171
— — — — minor	171
— — femoris	262
— retractor ani	235
— — anguli oris	115
— — costæ	220
— — labii, inferioris	116
— — — superioris	116
— — lingualis	134
— — penis	244
— retrahentes aurum	166
— rhomboideus	183
— sartorius	260
— scalenus	178
— scapulo-humeralis externus	203
— — — posticus	205
— — -ulnaris	205
— scuto-auricularis externus	122

	PAGE		PAGE
Muscles, scuto-auricularis internus	122	Mycloplaxes	8
		Myolemma, *see* Sarcolemma.	
— semimembranosus	259	Muzzle	24
— semispinalis dorsi, &c.	186		
— semitendinosus	259	Naboth's glands	250
— serratus magnus	218	Nares, anterior	139
— — parvus anterior	184	— posterior	138
— — — posterior	183	Nasal cartilages	139
— sphincter ani	235	— chambers	139
— — vesicæ	242	— fossæ or cavities	139
— spinalis dorsi	185	— mucous membrane	145
— — colli, &c.	179	— nerve	145
— splenius	170	— openings	139
— sterno-maxillaris	167	— peak	33
— — -thyro-hyoideus	167	Nates	158
— stapedius	165	Navicular fossa	243
— stylo-maxillaris	168	— — female	257
— — -pharyngeus	139	Neck, dissection of	166
— — -hyoideus	136	— of tooth	128
— subscapularis	204	Nerve	13
— subscapulo-hyoideus	217, 200	— cells	14
— superior oblique, of eye	144	— fibres	14
— — of tail	272	— tissues	13
— temporalis	124	Nerve, abducens	145
— tensor palati	137	— auditory	164
— — tympani	165	— auricular	182
— — vaginæ femoris	258	— axillary plexus	207
— teres externus *v.* minor	203	— brachial, *see* Axillary.	
— — internus *v.* major	204	— bronchial plexus	192
— thyro-arytenoideus anticus	176	— buccal (of fifth)	119
		— cardiac	195
— — — posticus	176	— cervical	182
— — -hyoideus	176	— chorda tympani	124
— — -pharyngeus	138	— coccygeal	272
— of tongue	134	— crural	266
— trachelo-mastoideus	170	— dental, superior	127
— trachealis transversus	177	— diaphragmatic, *see* Phrenic.	
— transversalis abdominis	221	— eighth cranial	164
— — costarum	184	— eleventh cranial	163
— trapezius	183	— facial	119, 126, 148
— triceps abductor femoris	258	— fifth cranial	127
— — extensor brachii	203	— first cranial	148
— — — cruralis	262	— fourth cranial	145
— ulnaris accessorius	210	— glosso-pharyngeal	139
— vastus externus	262	— gluteal	266
— — internus	262	— gustatory	124
— Wilson's	242	— hypoglossal	136
— zygomaticus	115	— intercostal	404
Musculi pectinati	197	— laryngeal, superior	177
— papillares	198	— — inferior	177
Musculus ciliaris *v.* ligament	148	— lingual	136

INDEX.

	PAGE
Nerve, lumbar	265
— maxillary, superior	127
— — inferior	124
— metacarpals	287
— metatarsals	272
— motores oculorum	145
— nasal, lateral	145
— ninth	139
— obturator	266
— olfactory	141
— ophthalmic, of fifth	145
— optic	148
— palato-maxillary	127
— pathetic	145
— phrenic	178
— plantar	287
— pneumogastric	177
— popliteal	271
— portio dura	164
— portio mollis	164
— radial	207
— recurrent	177
— sacral	266
— scapular	207
— sciatic	266
— seventh	164
— sixth	145
— spinal accessory	182
— spiral	207
— splanchnic	195
— subscapular	207
— suboccipital	182
— supra-orbital	145
— sympathetic	195
— third cranial	145
— tibial, anterior	271
— — posterior	272
— twelfth	136
— ulnar	207
— vertebral	195
— vestibular	164
— vidian	127
Nervosa, tunica	148
Nervus vagus, *see* Pneumogastric.	
Neural arch	150
Neurilemma	14
Nipples, *see* Teat.	
Non-articulatory eminences	9
Non-ciliated epithelium	16
Nostril, false	124
Notch	10

	PAGE
Notch, intervertebral, *see* Gap	143
— ischiatic	110
Nuchæ, ligamentum	104
Oblique muscles of abdomen	220
— — eye	144
— flexor of metacarpus	211
Oblongata, medulla	159
Obturator foramen	84
Occiput	24
Occipital sinuses	154
Odontoid ligaments	103
— process	49
Odoriferous glands	244
Œsophagus	172
Olecranon	71
Olfactory bulb	160
— fossa	30
— nerve	160
— sinuses	150
Olivary bodies	159
Omenta, gastric	223
Openings, pupillary	148
Ophthalmic ganglion	145
Optic decussation	160
— disc	48
— fossa	29
— thalamus	157
Ora serrata	147
Orbiculare, os	164
Orbital process of os frontis	27
Organ	1
Os, *see* Bone.	
— planum	28
— uterinum externum	248
— — internum	249
Ossicula auditûs	164
Ossific tentorium	26
Ossification	7
— centres of	7
Osteology, descriptive	24
Otic ganglion	124, 165
Otoconites	163
Otolithes	163
Ovale, foramen	254
Ovaries	248
— Graafian vesicles of	248
— ligaments of	249
— tunica albuginea of	248
— structure	248
— artery of	238

	PAGE		PAGE
Ovum	248	Peptic glands	230
		Perforated spaces of brain	161
Pacchioni, glands of	154	Pericardium, fibrous	192
Palate, hard	128	— serous	193
— soft	128	Pericardial sac	193
Palatine arch	38	Perichondrium	105
— canal	36	Perilymph	163
Palpebral fissure	121	Perineum	241
Pampiniforme, corpus	246	Peri-orbitale	143
Pancreas	224	Periosteum	9
Pancreatic duct	224	Peristaltic action of bowels	232
Panniculus carnosus	183, 216	Peritoneum	222
Papillæ	16	Permanent cartilage	7
— dental	130	— teeth	128
— hair-follicles	17	Pes anserinus	127
— skin	17	Petit's canal	147
— tongue	133	Peyer's patches	233
— of bowels, see Villi	232	Pharynx	138
Papillary layer of skin	17	Pia mater of brain	155
— stage of tooth development	130	— — spinal cord	180
		Pigmentary layer of choroid	147
Par magnum, see Nerve, pneumogastric.		Pillars of the fauces	137
		— — tongue	132
Parietal ridges	26	Pincers	129
Parotid gland	167	Pineal body	158
— duct	168	Pituitary body	154
Pars trigemini	126	— membrane	140
Partially moveable joints	100	Planum, os	28
Patches, Peyer's	233	Plates, cribriform	30
Patella, straight ligaments of	112	Pleura	189
Pathetic nerve	145	Plexus	14
Pectinean tubercle	85	— axillary	218
Pectiniforme, septum	243	— choroides, lateral ventricle	155
Pedicles	42	— — fourth ventricle	155
Pellucid zone	248	— — hypogastric	227
Pelvis, ligaments of	109	— — solar	227
— of kidney	240	Pneumogastric nerve	177
Pendulous portion of soft palate	138	Polar nerve-cells	14
		Pons Varolii	159
Penis	243	Porta of liver	225
— bulb of	244	Portæ vena	226
— corona glandis of	244	Portio dura	164
— corpus cavernosum of	243	— mollis	164
— — spongiosum of	244	Porus opticus	146
— crura of	243	Pouches, guttural	137
— glans of	244	Poupart's ligament	219
— ligaments of	245	Præputium clitoridis	252
— meatus urinarius of	243	— penis	244
— muscles of	244	Premolar teeth	132
— septum pectiniforme of	243	Prerenal capsule	236
Penniform muscle	12	Preservation of subject	4

INDEX.

	PAGE
Primitive dental groove	130
Process	10
— auditory, external	32
— basilar, of occiput	25
— coracoid, of inferior maxilla	40
— — scapula	61
— crista galli	29
— hamular, of ulna	77
— hyoid	32
— mastoid, petrous temporal	32
— — squamous temporal	31
— oblique	43
— odontoid, of axis	49
— olecranon, of ulna	71
— palatine, of palatine bone	37
— — of anterior maxilla	34
— — of superior maxilla	36
— styloid occiput	25
— — petrous temporal	32
— spinous, anterior inferior, of ilium	82
— — superior posterior of ilium	83
— — inferior	43
— — superior	42
— transverse	43
— vermiform	159
— zygomatic	31
Processes, ciliary	147
Prostate gland	243
Prostatic portion of the urethra	242
Pterygoid ligament	38
Pubio-femoral ligament	111, 261
Pulp of tooth	129
Puncta lachrymalia	121, 143
— vasculosa	156
Pupil of the eye	148
Pupillary opening	148
Pylorus	229
Pyramids of medulla oblongata	160
— kidney	240
Quadrigemina, corpus	158
Racemose glands	19
Radialis accessorius	210
Radio-carpal joint	107
— -ulnar arch	69
— — articulation	106

	PAGE
Ramus anastomoticus	168
— of inferior maxilla	41
Raphé corporis callosi	156
— scroti	244
Receptaculum chyli	227
Recti muscles of eye	144
Recto-vesical folds	241
Rectum	235
Rectus abdominis	220
Red muscular fibre	12
Renal capsule	239
Reproductive organs, see Generative organs.	
Restiform bodies	160
Rete mirabile	20
— mucosum	17
Reticulated layer of skin	17
Retina	148
Retractor ani	235
— costal	220
— penis	244
Ribs	56
— false	59
— first	59
— floating	58
— last	60
— true	58
Ridge	10
— alveolar	42
— mastoid	25
Rima glottidis	176
Ring, abdominal, external	219
— — internal	219
— vertebral	43
Root of the lung	191
Rotatorial joint	101
Round ligament of bladder	222
— — liver	223
— — ovary	224
— — uterus	224
Ruyschiana tunica	147
Saccular stage of tooth development	130
Saccules	163
Sacs, air	191
Saggital suture	102
Salivary glands	118
Sarcolemma	20
Sarcous elements	12
Satellites	12

INDEX.

	PAGE		PAGE
Scabrous pits of occiput	26	Solar plexus	227
Scalæ of cochlea	163	Sole, horny	276
Schindylesis	99	— sensitive	279
Schneiderian membrane	141	Space, femoral	260
Scissatus, lobus	225	Spermatic cord	247
Sclerotic	146	Sphenoideal sinus	142
Scrotum	245	Spheroidal epithelium	16
Scutiform cartilages	122	Sphincter ani	235
Sebaceous glands	18	— vesical	242
Sella turcica	30	Spigelii, lobus	225
Semicircular canals	163	Spinal cord	180
Semilunar cartilages of stifle	111	— column	42
— ganglion	227	Spines, *see* Processes, spinous.	
Seminal ducts	241	Splanchnic nerves	195
— vesicles	241	Spleen	235
Seminiferous tubes	241	— arteries	236
Sensitive frog of foot	280	— capsule	236
— sole	279	— curvatures	236
— structures	278	— hilum	236
Sensory roots of nerves	181	— internal structure	236
— nerve fibres	13	— pulp	236
Septum lucidum	156	Splint bones	79
— nasi	139	— hind	98
— pectiniforme	243	Spongiosum, corpus	244
— scroti	245	Spongy urethra	243
Serous membrane	15	Squamous epithelium, *see* Tesselated.	
Sheaths of tendons	15		
Sheath, medullary	14	— suture	99
Short bones	9	Stages of tooth development	130
Shoulder-joint	105	Stapes	164
Simple glands	18	Stenson's duct	141
Sinus	10	Sterno-costal cartilages	58
— of auricle	197	Stifle-joint	111
— of dura mater	153	Stomach	229
— olfactory	150	Straight patellar ligaments	112
— Valsalvæ, aortic	199	Stratified epithelium	16
— — pulmonary	199	Striæ	12
Sinuses of head	142	Striated muscular fibre	11
— venous	22	Subarachnoid fluid	155
Skeleton, artificial	7	— space (of cord)	180
— natural	7	— — (of brain)	155
Skin, dermis	17	Subcarpal ligament	210
— epidermis	17	Subcutaneous areolar tissue	17
— glands of	17	Subject, choice of a	3
— papillated layer of	17	Substantia perforata	160
— rete mucosum	17	Sudoriparous glands	18
— incisors through the	4	Suet	239
Skull, bones of	24	Sulci	155
Slit of the external ear	123	Superficial abdominal fascia	216
Small intestines	233	Superior maxillary sinus	142
Smegma præputii.	244	Supports for parts of subject	6

INDEX. 309

Supra-renal bodies, *see* Pre-renal bodies.
Suspensory ligament, bladder . 223
— — liver . . . 223
— — penis . . . 216
Suture 99
Sutures of skull . . . 102
Sylvius, fissure of . . 161
— aqueduct of . . . 158
Sympathetic gangliated cord 196
— nervous system . . 13
Symphysis maxillæ inferioris 141
— ischii 84
— pubis 85
Synarthrosis . . . 99
Synovial fringes . . . 15
— membranes . . . 15
— joints 100

Tænia semicircularis . . 157
Tail 272
Tapetum lucidum . . 149
Tarsal cartilages . . . 120
Tarsus 92
Teeth 128
— canine 129
— cavities of reserve of . 130
— cementum . . . 129
— compound . . . 129
— crown of . . . 128
— crusta petrosa of . 129
— dentine of . . . 129
— development of . . 130
— enamel of . . . 129
— incisor 129
— infundibulum of . . 12
— milk 129
— molar 131
— primitive dental groove . 130
— pulp, dental . . . 129
— table of development of . 121
— temporary . . . 128
— — eruptive stage . 130
— — follicular stage . 130
— — papillary stage . 130
— — saccular stage . . 130
— — shedding of . . 130
— wolf's 132
Tendons 12
Tensor tympani . . . 165
Tentorium, membranous . 153

Tentorium, ossific . . 26
Teres ligament . . . 111
Testes cerebri . . . 158
— (testicles) . . . 240
— coni vasculosi . . . 246
— coverings of . . . 245
— descent of . . . 246
— epididymis . . . 247
— globus anterior . . 247
— — posterior . . . 249
— gubernaculum . . 246
— lobuli 246
— mediastinum . . . 246
— tunica albuginea . . 246
— — vaginalis propria . 246
— — — reflexa . . 246
— — vasculosa . . . 246
— vas deferens . . . 247
Thalami optici . . . 157
Thebesius, foramina of . . 197
— valve of 197
Thecal 15
Thorax 188
Thymus gland . . . 200
Thyroid body . . . 166
— cartilage . . . 173
Tissue, yellow elastic . . 7
— white fibrous . . . 7
Tongue 132
— bone of 133
— frænum 132
— mucous membrane of . 132
— muscles of, extrinsic . 134
— — intrinsic . . . 134
— nerves of . . . 136
— papillæ of . . . 133
— — circumvallate . . 133
— — filiform . . . 133
— — fungiform . . 133
Torcular Herophili . . 154
Trabeculæ of corpus cavernosum 243
— of spleen . . . 236
Trachea . . . 176, 191
Tract, optic 160
Transverse fissure of brain . 155
— — liver . . . 225
Tricuspid valve . . . 198
Trochanter major . . . 87
— minor, external . . 86
— — internal . . . 86

INDEX.

	PAGE
Trochanteric fossa	86
Tube, Eustachian	138
Tuber cinereum	161
Tubercle of atlas	47
— lachrymal	36
— of rib	57
— pectinean	85
— superior maxillæ	35
Tuberculum Loweri	197
Tuberosity	10
— of ischium	85
Tubes, Fallopian	249
— uriniferous	240
— contorted	246
— seminiferous	246
Tunica adiposa renalis	239
— albuginea testis	246
— Jacobi	148
— nervosa	148
— Ruyschiana	147
— vaginalis scroti	224
— — testis	224
— vasculosa retinæ	148
— — testis	246
Turbinated bones	33
Tushes	29, 132
Tympanic cavity	164
— — ossicula of	164
Umbilical artery	253
— cord	253
— region of abdomen	231
— vein	253
Unstriated muscular fibre	11
Urachus	255
Ureters	240
Urethra (male)	242
— fossa navicularis	243
— lacunæ of	242
— mucous membrane of	242
— membranous	242
— prostatic	242
— spongy	243
Urinary bladder	241
Uriniferous tubes	240
Uterus	249
— artery of	238
— body of	249
— cavity of	249
— cervix of	249
— cornua of	249

	PAGE
Uterus, epithelium of	250
— fundus of	249
— ligaments of	249
— mucous membrane of	250
— os or mouth of	249
Utriculus	163
Uvea	147
Vagina	250
Vaginal cavities	249
— portal veins	226
— tunics	224
Vagino-rectal fold	251
— vesical fold	251
Valsalva, sinus of	199
Valve, bicuspid	199
— Eustachian	197
— ileo-cæcal	233
— mitral	199
— semilunar aortic	199
— — pulmonary	198
— of Thebesius	197
— tricuspid	198
— of veins	22
— of Vieussens	158
Varolii, pons	159
Vas deferens	247
Vasa afferentia	23
— efferentia	23
— recta	246
— vasorum	20
Vascular coronary secreting substance	278
— laminæ	279
Vasculosa, tunica	148
Vegetative nervous system	13
Veins	21
— valves of	21
— axillary	207
— azygos	193
— brachial, superficial	207
— bronchial	192
— cardiac	200
— cava anterior	193
— — posterior	193, 238
— cephalic	207
— coronary	200
— circumflex, of ilium	265
— diaphragmatic, see Phrenic.	
— femoral	265
— of foot	215

INDEX. 311

	PAGE		PAGE
Veins, hepatic	226	Vertebræ, lumbar	52
— humeral	207	— sacral	54
— iliac, common	265	— true	43
— — external	265	— typical	42
— — internal	265	Vertebral segments	42
— intercostal	187	— articulations	103
— interlobular	226	— ligaments	103
— intralobular	226	— column, see Spinal column.	
— jugular	172	Vena montanum	242
— lingual	136	Vesica urinaria, see Bladder.	
— lumbar	239	Vesicles, germinal	248
— mesenteric, anterior	228	— Graafian	248
— posterior	228	— seminal	247
— metacarpal	215	Vestibule	162
— metatarsal	271	Vidian nerve	127
— phrenic	239	Vieussens, valve of	158
— plat	207	Villi of bowels	232
— portal	226	Villous layer of stomach	230
— pulmonary	198	Vitreous humour	149
— renal	239	— tables	150
— saphenous	60	Vocal cords	175
— spermatic	239	Voluntary muscle	125
— spiral	207	Vomer	39
— splenic	236	Vulva	251
— sublobular	226	— commissures of	251
— tibial	271	— erectile tissue of	251
— vaginal, portal	226	— fossa navicularis of	251
— varicose, of face	118	— labia of	251
Velum pendulum palati	138	— mucous membrane of	251
Vena cava, anterior	193		
— — posterior	193, 238	Wall of hoof	274
— portæ	226	"Water bag"	255
Venæ Galeni	153	Wharton's duct	132
— vorticosæ	147	White muscular fibre	11
Venous sinuses	22	— nerve fibres	14
Venter scapulæ	63	— substance of Schivann	14
— ilii	84	Wilson's muscle	242
Ventricles of larynx	176	Winslow's foramen	223, 230
Ventricle, fifth	156	Womb, see Uterus.	
— fourth	158	Wrist, see Carpus.	
— lateral	156		
— left (heart)	199	Xiphoid cartilage	60
— right —	198	Zima, zonula of	144
— third	158	Zona pellucida	248
Vertebræ, cervical	46	Zonula cilaris (v Zinnii)	149
— coccygeal	56	Zygoma	31
— dorsal	50	Zygomatic ridge	35
— false	44	— bone	38

ERRATA.

Page 20.—Line 42, *for* circular *read* circulus.
" 34.— " 4, *for* incisorium *read* incisorum.
" 41.— " 12, *for* carinus *read* caninus.
" 59.—Lines 41 and 44, *for* scalenius *read* scalenus.
" 76.—Line 33, *for* unciform *read* unciforme.
" 80.— " 38, *for* sessamoid *read* sesamoid.
" 108.— " 21, *for* suffraginis, *read* coronæ.
" 129.— " 12, *for* infundibuli *read* infundibula.
" 178.— " 27, *for* scalenius *read* scalenus.

www.ingramcontent.com/pod-product-compliance
Lightning Source LLC
Chambersburg PA
CBHW030748230426
43667CB00007B/883